Fossil Vertebrates
of Alabama

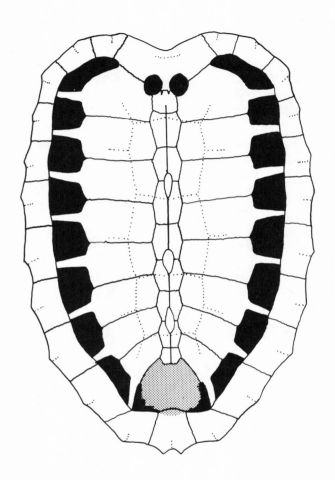

Fossil Vertebrates of Alabama

John T. Thurmond

AND

Douglas E. Jones

The University of Alabama Press
University, Alabama

Library of Congress Cataloging in Publication Data

Thurmond, John T.   1941–
  Fossil vertebrates of Alabama.

  Bibliography: p.
  Includes index.
  1. Vertebrates, Fossil. 2. Paleontology—Alabama.
I. Jones, Douglas E., joint author. II. Title.
QE841.T48   566′.09761   80-13075
ISBN 0-8173-0006-6

Contents

Acknowledgments

Many people contribute to the organization of a book of this type, especially one that has been so long in the making. During the time of the preparation of the manuscript, some who contributed to its composition have gone on to other endeavors. Mr. William E. Marsalis, Jr., who worked with Jones in the early days in the field, now is a geologist in private industry. Other graduate students, too numerous to mention, who were involved in various excavations are now at teaching and industrial jobs throughout the country.

The manuscript has required a number of revisions and retypings over the years. Our sincere thanks go to Mrs. Thomas A. Simpson and Elizabeth A. Netemeyer for their careful attention to a complicated manuscript and bibliography.

Special appreciation is due the Geological Survey of Alabama for its initial support of field investigations and early manuscript preparation. Mr. Charles W. Copeland, Jr., of that organization was particularly helpful in advising about the manuscript in the early part of the enterprise. Professor Forrest McDonald of the Department of History at The University of Alabama made useful suggestions regarding the introductory material. We appreciate the permission of the Paleontological Research Institution to use illustrations from one of its publications (White, 1956).

To all workers who shared information about collecting sites or supplied specimens themselves, we are especially grateful. While we acknowledge the help of a number of people, the authors of course assume responsibility for the information presented and the conclusions reached.

Preface

Perhaps no field of geologic endeavor fires the public imagination so readily as the study of fossil vertebrates. Visions of prehistoric monsters combine with a desire to know more about the antecedents of familiar animals and of human beings as well. As a result, the subject of vertebrate paleontology has made a surprising rise in popularity—surprising because economic applications of such studies are extremely rare, at least in any direct and obvious sense.

Much of the lack of economic uses of vertebrate paleontology lies in poor communication. Study of vertebrate fossils sometimes can provide answers for crucial economic questions that no other method can attack. In recent years considerable effort has been spent in an attempt to understand the precise conditions under which rocks were formed. Such a study has many facets, depending on the type of rock involved, but an important aspect is the examination of the rock's fossil content for information about the environments in which the animals might have lived. Many rock units in which vertebrate studies could have been significant have not yet been investigated. Why? Vertebrate paleontology exists on the fringes of two great areas of science—geology and zoology. As a result, workers in the field generally have acquired a technical vocabulary in both areas—half of which may be largely alien to a geologist or a zoologist. Consequently, published data in vertebrate paleontology tend to be ignored by workers in related fields, even in cases where such data would be useful in their own work. The breakdown of this barrier to the flow of knowledge will require considerable effort by all concerned.

Why has there been such interest in paleoecology, the study of the conditions under which rocks were formed? Of the two major applications of this study, one is immediate and pressing; the other is more remote but may be required to ensure the survival of life on the only planet we have. First, the more we understand the conditions of rock formation, the better we can predict the distribution of rock types, and thus delimit the most favorable areas for the development of their contained mineral resources. The never-ending search for oil and gas is a classic example. Roughly one out of eight wildcat wells drilled in this country produces oil. This ratio has remained remarkably constant from year to year throughout this century, despite the ever-dwindling number of unexplored areas considered favorable. Only our constantly increasing knowledge of rocks and their relationships to each other has kept this success/failure ratio from a downward spiral.

On a more distant plane, paleoecology is useful in determining the natural variations in conditions to which the earth is subject. We know now that human history is much too brief to have experienced all the possible disasters and blessings accorded this planet. For example, we have not seen what an ice age could do to civilization, or how a major advance of the sea would affect our world. We do not even know the causes of either of these events, but we do know that both have occurred many times in the history of the earth. This body of evidence is being added to and evaluated. The time may not be far distant when the survival of large parts of the human population will rely on our knowledge of natural phenomena.

These concerns and the realization that Alabama's significant fossil-vertebrate fauna is poorly known gave rise to this book. It is our hope that this effort will prove useful and valuable to those who derive pleasure, avocational or vocational, from the study of fossil vertebrates.

In 1964 Jones began this work with the late Ralph D. Chermock, then a professor of biology at The University of Alabama. Materials collected over a hundred-year period and in the holdings of The University of Alabama Museum of Natural History were organized and new collections made where possible. The original manuscript, very incomplete, lay dormant for about five years after Chermock left the University, even though some limited field work was continued by Jones until 1968 when he became Dean of the College of Arts and Sciences. The original concept of a study of Alabama fossil vertebrates, the inspiration of Ralph Chermock, probably would have died had not the current authors become acquainted in 1971. Thurmond revised and rewrote much of the manuscript, bringing to the enterprise a high degree of professional expertise in vertebrate paleontology. At the time Thurmond was on the faculty at Birmingham-Southern College and the author of a number of publications on fossil vertebrates in the Gulf Coast region. Joint efforts to bring this work to a close (one never completes such a study) have continued since 1971 even though both authors have had to deal with the competition of teaching and administrative duties in the interim. The contribution on fossil reptiles by Samuel W. Shannon (Appendix I) is evidence of the continuing work in Alabama. Notably, Cailup B. Curren, Jr., proceeds with research on Alabama's Pleistocene mammals.

In the perspective of the enormous length of geologic time encompassed in the history of fossil vertebrates in Alabama, the years involved in preparing this book are indeed insignificant. However, mere mortals exist within a more restricted time frame and must consider the needs and expectations of the scientific community to have the information contained herein provided in a timely manner. For our efforts to be meaningful to professionals and amateurs alike, especially in light of the constant changes in scientific nomenclature, the results of our work must be published and subjected to scrutiny. So be it!

Vertebrate Paleontology

Introduction

Vertebrate Paleontology as a Science

WANTED—individual with vast knowledge of geology and zoology, not afraid of hard manual labor, both summer and winter. Considerable ability as promoter or sales manager essential. Other factors considered will be ability to operate heavy equipment, training as stonecutter or sculptor, and reading knowledge of several languages (French, German, Spanish, Portugese, Italian, and Russian). Must be willing to work long hours for satisfaction, as financial rewards not large.

How many responses to such an ad would you expect? Any one individual accomplishing all the work expected in vertebrate paleontology needs to have all these talents, along with several others. Only a very few individuals have such a combination. Fortunately, work may be shared by several persons of varying areas of specialization, so cooperation usually makes progress possible.

As a result of this needed combination, work in vertebrate paleontology is carried on at comparatively few centers in this country. Major centers include the American Museum of Natural History in New York, the Museum of Comparative Zoology at Harvard, the United States National Museum in Washington, the University of Kansas, and the University of California. Only one such major center exists in the southeastern United States and is a broadly based group of workers at the Florida State Museum, University of Florida.

In addition to these major centers, there are a number of minor ones where individuals or small groups are working. Their efforts are often of great significance in their areas of specialization. Such smaller centers usually are associated with college or university departments of geology or biology or with university museums. The amount of work accomplished by these small groups is amazing because they are almost always part-time efforts by interested professors on limited budgets. Some are short-lived, depending on the presence of a single worker, while others show a remarkable continuity. In general, the southeastern United States lacked such small centers until recently. Several programs have begun since 1965 and now show considerable promise of sustained work.

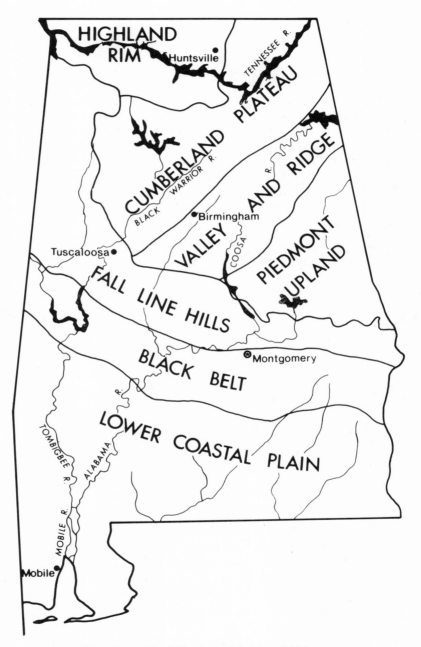

FIG. 1A.   Generalized Physiographic Map of Alabama.

Distribution of Alabama Vertebrate Fossils

Alabama occupies an important place in North American geology and paleontology. Much significant early work in American Cretaceous and Tertiary geology has taken place in the state since the early 1800s. Many stratigraphic names widely employed in the Gulf and Atlantic Coastal Plain Province have their origin in outcrops along Alabama roadsides and rivers. Early work in Alabama by Isaac Lea and T. A. Conrad, among others, set the stage for much of the systematic paleontology in use today in the Cenozoic around the world. Great commonality also exists between depositional environments of some Paleozoic strata in Alabama and places as distant as the northern Appalachians and the British Isles.

Consequently, Alabama is not an isolated Southern state with a provincial assemblage of fossil vertebrates. It is a place that records in its geologic framework faunal and depositional conditions found in many other places in the world. The state is unique, perhaps, in that many of the Cretaceous and Tertiary strata are highly fossiliferous and beautifully exposed. Various workers consider, for example, the exposures of marine Tertiary in southwest Alabama to be the finest in the world.

Many of the forms discussed in this book are to be found in beds of equivalent age in other parts of the United States or the world. The techniques of collection and curation presented herein are cosmopolitan in application as are the principles of systematics. It may be possible for a reader to identify a Kansas mosasaur by use of descriptive and illustrative material concerning a mosasaur excavated in Perry County, Alabama. Geology recognizes no political boundaries.

Vertebrate fossils have been found in almost every corner of the state, from Tuscumbia to Mobile to Dothan. The greatest abundance of material has come from the Black Belt and regions farther south (fig. 1A). Other regions have received so little work that the scarcity of remains may be more apparent than real. Only the intensely deformed rocks of the Piedmont metamorphic belt, roughly the east-central part of the state, should be considered unfavorable. Even here, there is opportunity to find comparatively late material in superficial deposits, as along streams. No person in Alabama is out of reach of fossil vertebrates, and almost every backyard is a potential site.

The simplest way to review the known fossil vertebrates of Alabama is by age. For the relationship of the age terms and the distribution of rocks of these ages, see the state geologic map (fig. 1B). Each type of animal will be described in more detail later.

No vertebrate fossils are known from the Cambrian, Ordovician, or Silurian rocks of Alabama. The Cambrian predates the appearance of

vertebrates (except for the debatable conodonts). Ordovician vertebrates are rarest and known only from fragments. The chance of finding one is small, but a lucky find would certainly ensure the dis-
coverer a place in the history of paleontology. The Silurian rocks of
Alabama are more promising, especially in the Red Mountain Formation, but no results have been obtained yet.

Devonian rocks are known to bear rich faunas of fishes in other parts
of the world, but our knowledge of Alabama vertebrates begins at the
very end of the period. Only two specimens are known, from rocks
that lie at the boundary between the Devonian and the Mississippian.
No new material has been reported since the description of these two
specimens by Tuomey (1858A). Our knowledge of the Mississippian
vertebrates is also based on a few century-old specimens, though a
new find is noted below (see *Petalodus*). Pennsylvanian vertebrates
are known only from tracks in Alabama, as reported by Aldrich and
Jones (1930).

GENERALIZED GEOLOGIC COLUMN OF ALABAMA

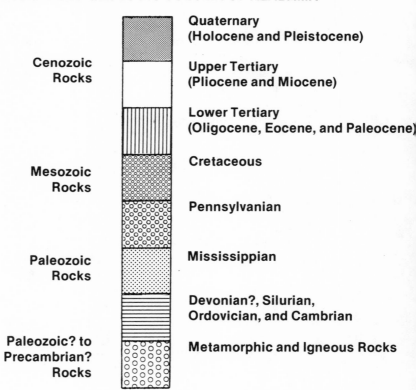

Cenozoic Rocks — Quaternary (Holocene and Pleistocene)

Upper Tertiary (Pliocene and Miocene)

Lower Tertiary (Oligocene, Eocene, and Paleocene)

Mesozoic Rocks — Cretaceous

Pennsylvanian

Paleozoic Rocks — Mississippian

Devonian?, Silurian, Ordovician, and Cambrian

Paleozoic? to Precambrian? Rocks — Metamorphic and Igneous Rocks

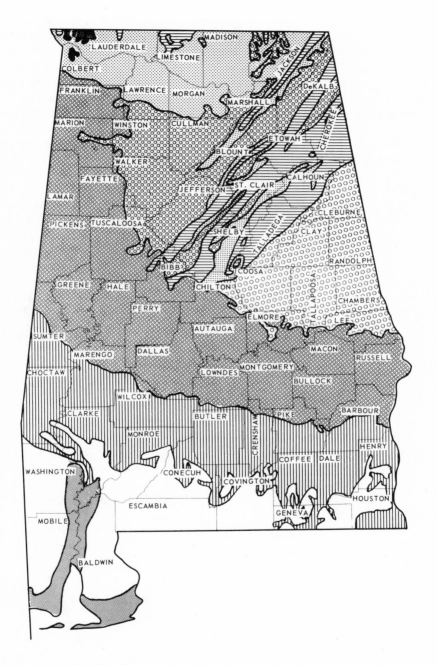

FIG. 1B.   Generalized geologic map of Alabama. Geological Survey of Alabama, Circular 47, pl. 1, 1968.

No rocks of the Permian, the last period of the Paleozoic, are known to crop out in Alabama. Also, no outcrops are evident from the first two periods of the Mesozoic, the Triassic and the Jurassic.

Our knowledge of Alabama vertebrate fossils begins again with the Upper Cretaceous. No material has been reported yet from the Tuscaloosa Group, the basal part of this series, so extensive work is needed. The character of the rocks is very similar to that of deposits in other areas known to yield rich faunas of shallow-marine, freshwater, and land vertebrates. Most likely the material here will not be conspicuous; small teeth and scales probably will be the only visible remains. Very little is known of the fauna of the Eutaw Formation overlying the Tuscaloosa. A mosasaur has been collected from this unit near Columbus, Mississippi (Cope, 1869A; Russell, 1967), and Applegate (1970) reports several fishes from either the top of the Eutaw or the basal rocks of the overlying Selma Group.

The most thoroughly known fossil-vertebrate fauna of Alabama is in the Selma Group, the chalky rocks forming the Black Belt. Our knowledge of this fauna is summarized in a magnificent series of papers from the Field Museum of Natural History (Zangerl, 1948A, 1948B, 1953A, 1953B, 1960; Langston, 1960; Russell, 1970; Applegate, 1970). Very little is known of material from the younger Cretaceous rocks.

The Paleocene rocks of Alabama hardly have been touched, although Leriche (1942B) notes fishes from the Midway Group of Wilcox County. Nothing is known from the lower Eocene Wilcox Group at this time. White (1956) describes fishes from the Claiborne Group, while recent work (Thurmond, in prep.) indicates that this unit is extremely rich in fishes, reptiles, and mammals.

The uppermost Eocene rocks, the Jackson Group, have been studied extensively, yet considerable work remains. These rocks are the home source of famous Alabama basilosaurs and other whales. (See discussion at Cetacea, below.) Lucas (1898B) describes a sea snake from these rocks, and Woodward (1889D), Leriche (1942B), and White (1956) discuss the fishes.

There seems to be no published information on Oligocene or Miocene vertebrates from Alabama, though the Oligocene Red Bluff Formation has produced at least one specimen. A site of probable Pliocene age discovered near Mobile in the late 1960s resulted in the publication of a preliminary faunal list of this site by Isphording and Lamb in 1971.

Pleistocene vertebrates have been recorded from scattered sites all over the state, but to date (1980) no locality has been worked extensively and reported in the literature. Findings are certainly only a fraction of the material available. Early investigators (Leidy, 1855B,

p. 6; Tuomey, 1858A; Mercer, 1897) reported giant ground sloth and mastodon from caves near Tuscumbia. Falconer (1857B) included Alabama as one of the places where his new species *Elephas columbi* (Columbian mammoth) had been found, apparently basing his statement on a tooth from "Alabama, near the Gulf of Mexico" (Hay, 1923, pp. 164–165). From the central part of the state is evidence of mammoth, horse, and bison (Hay, 1923). These scattered finds only indicate the rich discoveries yet to be made.

Vertebrate Paleontology in Alabama

The initial discovery of a fossil vertebrate in Alabama involved the first good specimen of the gigantic Eocene whale, *Basilosaurus*, found in Clarke County about 1833. This animal had been described previously (Harlan, 1834C) from a single vertebra from Louisiana. Harlan (1835) described the much more complete Alabama specimen, and soon a lively transatlantic controversy was being waged over the exact relationships of the animal. As a result of this controversy, large quantities of bones from Clarke, Choctaw, and Washington counties were on the way to museums along the Atlantic Coast and in Europe. Many were pried loose from underneath houses, for which they served as foundation posts, to be utilized in their new role as ammunition in a spirited scientific debate. (Some basilosaur vertebrae still are used for house foundations in south Alabama—a rare case of a useful vertebrate fossil.)

Within a short time, this huge mammal became the subject of one of the wildest excursions of vertebrate paleontology into the world of showmanship. A peripatetic German scientist and promoter named Albert Koch diligently collected various parts of basilosaur anatomy throughout three counties in Alabama. He then assembled the entire collection, representing parts of at least six animals, into a vast "reconstruction," 110 feet long. This monstrous conglomeration was placed on display in the Apollo Saloon in New York (admittance 25 cents) and "scientifically" described in the program for the exhibit (Koch, 1845A) as *Hydrargos sillimanii*. The name of the genus means "sea chief," and the species was named in honor of the eminent Professor Silliman of Yale, founder of the *American Journal of Science*. This heartfelt (and commercial) dedication was soon scotched by Silliman himself, who insisted that the "species" be named after Harlan, who had first described the beast. Koch revised the program (Koch, 1845B) and no doubt preened himself for placing "credit where credit was due," while Silliman rejoiced over divorcing his name from such charlatanry. The learned Dr. Harlan was now the late Dr. Harlan and unable to defend himself.

After a triumphant (and profitable) exhibition in New York, Koch and the "*Hydrargos*" took ship for Europe where the monster was exhibited before the "crowned heads" beloved of the circus world. At least one of these displays produced a new program (Koch, 1846) in impressive scientific German. At some time before this latest program, the great beast had been rechristened *Hydrarchos* ("sea ruler"). The "skeleton" was finally sold for a large sum to the king of Prussia, who deposited it in the Royal Museum at Berlin. His scientists quickly dissected it, established that the large portion was identical to Harlan's *Basilosaurus*, and described several new species from the remaining fragments.

Koch was undaunted by the dissolution of his monster and used the proceeds of the sale to return to the collecting area in Alabama and assemble a new "*Hydrarchos*," said to have been even larger than its predecessor. This reconstruction was sent on a "Grand American Tour," changed hands several times, and finally perished in the great Chicago fire of 1871. It was justly so: this legendary beast of the past was destroyed by the equally legendary Mrs. O'Leary's cow.

Subsequent work on the vertebrate fossils of Alabama has been on a much firmer basis, but the élan of the old days has been lacking. From 1850 to 1870, there was a flurry of interest in the giant marine reptiles of the Cretaceous. Other work proceeded mostly in fits and starts. Specimens were found, shipped off to distant museums, and described when sufficiently important. Comparatively little work was based completely in Alabama, though what exists tends to be excellent.

A notable exception to the scattered work that has characterized Alabama vertebrate paleontology was the collecting expedition of the Field Museum of Natural History of Chicago in the late 1940s. This work was concentrated in the Selma Chalk of the Black Belt in western Alabama and produced a vast amount of material. Although several important and extensive papers already have resulted from this expedition, work is still going on (in 1980). As a result, the fauna of the Selma is known far better than that of any other unit in the state.

Alabama is rich in fossil vertebrates as attested by the geographic occurrences of genera and species found in the state as cited in later pages. Certain other areas of Alabama are likely ones for the discovery of fossil vertebrates. Because much of the material is very poorly known, there is excellent opportunity for both amateurs and professionals to make significant contributions in an area where work is badly needed.

The Amateur and the Professional

It is unfortunate that many on both sides of the fence would have entitled this section "The Amateur vs. the Professional." All too often, the two groups have worked at cross purposes or even with open hostility. Amateurs have been known to call professionals "greedy snobs who want to hog the fossils." Some professionals consider the amateur as a vandal whose principal activity is destroying good specimens.

There is no reason for this hostility beyond a failure of both groups to make their objectives known to each other. The ultimate objective is the same—to learn more about the life of the past. The differences lie in the intensity of the effort. Both the amateur and the professional may profit from cooperation instead of competition.

The main objective of amateurs is to increase personal knowledge and to gain pleasure in the process. They enjoy the excitement of the search, a hunt that may follow a trail, creek, or road never before traveled. The search gives a reason to stroll about outdoors with a feeling of doing something useful.

To the professionals, this activity is deadly serious. Professional reputations depend on the ability to produce new information to add to the ever-expanding province of the known. Personal knowledge increases in the process, but the objective is the growth of the world's knowledge of its own past. Discovery of a specimen is only the beginning of a professional's work.

These objectives color the attitude of each toward the specimens found. A vertebrate fossil collected by an amateur is a trophy of the chase. When identified, it gains even more interest, but it still is a proud trophy. To the professional, a specimen is a storehouse of information—something to be analyzed by every means possible. In some cases, the professional will even destroy it to gain further knowledge. Usually, however, after most professionals have exhausted their own abilities on the specimen, they will deposit it in some safe collection that is available to other workers. It may be studied later by many researchers, each trying to learn something new from it or to check information that is known already. Many specimens collected over a century ago, including some from Alabama, are still subjects of lively debate.

The true professional takes no pride in a personal collection other than in the work it represents. The fossils gathered do not belong to an individual. In a real sense, they belong to the world, including generations yet unborn. This attitude is the key to understanding the orientation of the professional paleontologist, who truly believes that an important specimen should not be hoarded in a private collection. Its proper place is in a museum where it may be preserved and still be available to anyone who needs to examine it. This feeling is one of two principal sources of friction between professional and amateur, for the latter has an understandable proprietary pride. There probably always will be friction over this point, but being aware of these differences can contribute to the understanding of both groups.

Another source of friction lies, oddly enough, in the good intentions of the amateurs, whose tendency is to underestimate the difficulty of collecting a vertebrate fossil. As a result, amateurs often inadvertently destroy magnificent specimens with improper collection methods. A good rule of thumb is this: *The better the specimen, the more difficult it will be to collect without damage.*

Where do the limits of the amateur's powers lie? It is impossible to answer the question simply. Each fossil is an individual as is each collector. When one individual meets another in this relationship, results are difficult to predict. Some cases are clear-cut. For example, shark teeth lying on the surface are obviously within anyone's capabilities. Complete skeletons certainly will require professional assistance for excavation. We will attempt later to help our readers evaluate the limit of their capability to handle a specimen in the field. Remember: *A ruined specimen is no good to anyone.*

Both the professional and the amateur have vital roles in the search for knowledge of the past. Each has advantages that are particularly suitable. The population of professional vertebrate paleontologists in this country is very small (though fairly dense by worldwide standards). Each paleontologist has a large stretch of territory to cover, far more than can be attended to in a lifetime, and spends much time finding fossils. If, however, the objective is to learn as much as possible about the life of the past, mere collecting is not the most efficient use of training and experience. Professional skills can be best utilized by gathering together specimens already known and studying them in the laboratory. Another important use for these skills is in prospecting remote areas where studies have indicated the likelihood of particularly important material.

The cooperation of the amateur is essential if the search for knowledge is to progress. By the most conservative estimate, there are at least a thousand amateurs for every professional paleontologist. Some would place this estimate as high as a hundred thousand to one. By sheer numbers, then, amateurs undoubtedly locate far more fossils each year than professionals do. But numbers do not tell all the tale. Most amateurs operate very near home, in territory they know well. They not only have their own experience but also that of past generations of local fossil hunters. The professional in a strange area is much less effective if he cannot, or will not, draw on the vast amount of local knowledge that is available.

Amateurs are as varied as people. They vary from the small child with a fossil shark tooth in a private box of treasures to a number of extremely competent amateurs with several published papers to their credit. The latter differ hardly, if at all, from professionals in their degree of competence. They are classed as amateurs only because this work remains a hobby rather than their life's work. In a number of cases, individuals have made the transition from amateur to professional. One of the authors studied under such a person.

Fossil Collection: Instructions and Help

Collecting vertebrate fossils is not a simple business. Proper collecting methods and techniques have not been much publicized at the amateur level. Those who have the knowledge and those who need it often have not communicated in a completely open manner. Professionals in the past have viewed the amateur largely with hostility, whether open or concealed; they have seen too many magnificent specimens destroyed by the well-intentioned, but misguided, attempts of amateurs. (Even more destructive are those rockhound addicts of dinosaur-bone jewelry; sometimes they will break up skeletons into polishable fragments.) Unfortunately, professional paleontologists have tended to react to these bitter experiences by retreating into secrecy. On the other hand, amateurs have been reluctant to seek professional help out of fear of losing their specimens. Each group must endeavor to establish mutual trust and cooperation if any good working relationship is to be developed.

Discovery; or, how to find a dinosaur

Fossil hunting is not unlike fishing in many ways. Fishermen who have made a careful study of the habits and habitats of fish are far more likely to be successful than novices, particularly on an unfamiliar lake. Fishermen will vary techniques according to the fish they are trying to catch; so will fossil hunters. Sometimes both will change methods because what they have been doing just has not produced results.

Where will hunters most likely find fossil vertebrates? The first rule is to look for fresh exposures (outcrops) of relatively unweathered rock. "Rock," as used here, is a relative term. It implies "unaltered material underlying the soil." The term may embrace anything from crumbly sands and clays to massive, solid rocks. Occasionally the hunter may find a fossil in the soil under heavy vegetation, just as a fisherman may catch a bass where no self-respecting bass should be. But the good fossil hunter, like the good fisherman, avoids spending much effort on such unfavorable areas.

Where can outcrops of rocks be found? They may be almost anywhere, but most are in two types of areas: where the earth is being rapidly attacked by erosion, and where it is being disturbed by human activity.

Most areas of erosion are along streams, both large and small. A field whose soil is being replaced by gullies can reveal many fossils. One professional hunter used to begin a visit to a new area by asking the local Soil Conservation Service officials to describe areas where they felt they had succeeded, where they were working actively, and where they felt the battle was lost before they began. He then concentrated his work in the latter.

Areas made suitable for exploration by human activity include road cuts, dam and building excavations, gravel pits, and openpit mines. *Such areas often are dangerous,* especially when active work is going on. Be sure to ask permission when there are signs of recent activity. No fossil is worth being caught in a rock fall or a quarry blast.

After locating an outcrop, the hunter must decide if it is worth investigating. With extremely rare exceptions, *only sedimentary rocks have fossils.* Vertebrate fossils, particularly, are seldom in other types of rocks. The sedimentary rocks include such soft materials as clay, sand, and gravels, and their hardened equivalents, shale, sandstone, conglomerate, and limestone. These are the only sedimentary rocks likely to be found in the field.

In general, gently sloping outcrops are to be preferred to vertical faces. A sloping face exposes more of each bed, offering a better chance of finding fossils if they are present. Besides, it won't fall in on you. Concentrations of fossils often are found at the contact of two different rock types. Beds of greenish sand or sandy clay usually are very rich, particularly in small specimens.

Lighting is a surprisingly important factor in a successful fossil hunt. Clear, bright days are best. The morning or late afternoon sun will create sharp shadows that make specimens stand out more clearly than at midday. Try to walk with the sun to one side; a sun in front is blinding, and from behind it puts your own shadow in the way. If the site is well known and frequented by collectors, try to be there as soon as it dries out after a good rain.

A very particular hunter who insists on looking only for dinosaurs should investigate two areas. Some dinosaur specimens have been found in the Selma Group of the Black Belt. These remains represent carcasses that floated out into the shallow sea in which the Selma was

deposited (Langston, 1960). The Tuscaloosa Group, forming the Fall
Line Hills, has produced nothing to date but appears promising.
Though chances are slim in such a search, the work needs doing.

Collecting; or, don't let that big one get away

Collecting vertebrate fossils ranges from picking up loose teeth on
the surface to an arduous and exacting exercise in patience. The prime
rule is: *Do not risk the specimen.* If you are not absolutely sure you
can extract the specimen without damage, do not try. Call in professional aid. If you try, and see that you are doing harm, stop and ask for
guidance. Destroying a specimen with the best of intentions does not
help anything, least of all your conscience.

Proper collecting for a given specimen is variable. Usually, large
bones are treated very differently from small bones and teeth. Collecting a large specimen is an extremely complex process. The general
principle is this: the bones are collected in a single lump, along with
their surrounding matrix. This block is carried back to the laboratory
and cleaned there (see below, under Preparation). The surface above
the skeleton is cleaned off to a foot above the bone or skeleton, using
whatever tools are most convenient. This work may involve anything
from a bulldozer to the ancient and honorable pick and shovel.
Paleontologists often regard the latter, especially the pick, as the
badge of the profession. The favorite type is the Marsh pick, a light,
one-handed pick with a two-foot handle, named after a well-known
nineteenth-century paleontologist. A specialized tool, it is rather expensive. An army-surplus entrenching pick is a good substitute (be
sure it says "U.S. Army," as imported substitutes are very brittle).
During this phase of "overburden removal," excessive vibration
should be avoided, though jackhammers and even dynamite have
been used on occasion. The secret of good work is to dig *down* to the
specimen, over an area wide enough to include all the bones. Working
in from the side increases the chances of cutting through a stray bone
that is sticking up. Also, you may produce a dangerous overhang that
can readily cave in.

The last foot of rock is removed by light hand tools only because
bone may be encountered at any time. Chisels, small picks, and even
dental tools are useful. When a bone is encountered, only enough of it
is uncovered to be sure of its extent. Any exposed surface is soaked
with a preservative. Very thin shellac is an old favorite, though some
of the more recent synthetic compounds have desirable properties.
Loose fragments are carefully glued back into place.

After the full extent of the specimen is known, a trench is run completely around it and should be carried down to about a foot below any

possible bone (deeper for a very large specimen). The bones now lie
on a block of rock surrounded by this trench. This block is undercut
along the bottom of the trench for several inches to a foot (again even
deeper for a very large block) all around its edge.

When the undercutting is complete, the top of the block is carefully
swept of all loose rock with a whiskbroom (a very useful tool throughout the dig), or with a paintbrush in very delicate places. All exposed
bone is then covered with wet paper, well tamped down to remove
any air bubbles. Any undercuts on the surface of the block should be
packed with wet paper.

With the block properly prepared, plastering may now begin. Cut
several burlap bags into strips about six inches wide. Make sure you
have enough strips to cover the block at least four times, with plenty to
spare. Most of the strips can be fairly short, but some should be as long
as possible. Put the long strips into a bucket of water, followed by the
short strips. Make sure all are thoroughly soaked. Mix a bucket of
plaster of Paris, starting with half or two-thirds of a bucket of water and
adding plaster slowly; mix well until the mixture is the consistency of
cream. If the specimen is large, mix enough plaster for the whole job.
Plaster won't stick to dry plaster!

Work quickly now. Kick over the bucket of water and burlap to let
some water drain out of the strips. Dip each strip in the plaster so that
it is completely covered. Apply the strips to the block, making sure
that each overlaps its neighbor for about two inches. Work in layers,
first one direction, then another. Tuck the strip ends well up under the
undercut block. After the first layer, some preselected tree limbs
and/or scrap lumber may be added for reinforcement. Finish off the
work with the long strips, running them around the block, packed into
the undercut.

You will now be ready for a well-earned rest. It is needed, too,
because of the plaster, which must set thoroughly before any more
work is continued. The time required varies from a few minutes to an
hour, but allow plenty of time. Then begin undercutting the block
again until it rests on a pedestal much smaller than itself.

The critical phase of the dig has now arrived. Take the point of a
shovel and drive it under the block as far as possible. Do this all
around the block. Using a pick or a shovel, pry the block *gently* all
around until it is clear that the block is completely free from the
underlying rock. Take a break to gather strength and nerve for the next
step.

As quickly as possible, turn the block over onto a level space. Don't
stop for anything, even if you hear material falling out of the bottom of
the block. Starting again will just make matters worse, so get it over.
Every now and then, a block will "swarm," break up and fall out of the

cast, when it is turned. It probably will not break if the previous work has been done properly, but this disaster has happened to almost every professional. Very wide and thin blocks are particularly likely to swarm, as are blocks in highly fractured material. Very rarely can a specimen that has swarmed be salvaged in anything like its original state.

With the block turned over, hollow it out as much as possible through the opening in the cast. This step reduces the weight that must be moved. If the block shows signs of breaking up, or if it must be moved across country for some distance, cut off the loose parts of the bottom (now the top) of the old cast, and apply a new cast to jacket the specimen completely in plaster bandages.

You are now left with a very rugged object that can take a lot of handling, provided it is not dropped, jolted, or turned on edge. Its weight may be anything from a few pounds up. Considerable ingenuity may be required to move a large block (occasionally a ton or more in weight) to a vehicle and to get it loaded. Paleontologists faced with this problem sometimes feel like Pharaoh's slaves working on a pyramid.

The whole procedure sounds like a lot of work. It is. Yet it is the only safe way known to collect the majority of fossil-vertebrate skeletons, and many isolated bones. The job is beyond the means of most amateurs and, in most cases, should be left to professionals. We fervently wish for a better way, but the basic procedure has not improved for a century or more.

Small specimens generally require much less work, unless you are conducting a large-scale search. Many fossils simply can be picked up. In a few places, splitting apart layers of rock can produce entire skeletons of fishes. Concretions (lumps formed from mineral matter carried in ground water) sometimes have fossils, even skeletons, in their centers.

Wet screening is a powerful tool in the search for many small fossils. The procedure can sometimes be complicated but consists essentially of digging up known fossil-bearing material, packing it in burlap bags, and allowing it to dry. The material is washed through screens (window screen is adequate for most purposes) and the residue carefully sorted for small bones and teeth. The mesh of the screen is crucial: we have seen people lose 90 percent of the specimens in the rock by using ¼-inch hardware cloth. Production can be extremely large in some places. One of the authors has been obtaining about 100 teeth per pound from some material.

In general, the proper collection of fossil vertebrates leads to greater returns, at a lesser cost in ruined specimens and frayed tempers. It is well worthwhile for both professional and amateur.

Preparation; or, you caught it, you clean it

There is perhaps no more frustrating task in the world than the preparation of a large fossil vertebrate. If there is, it is the preparation of a small fossil vertebrate in hard rock. Both are jobs requiring enormous amounts of patience and skill. The latter is useless without the former, and an exceptional dose of patience can make up for much lack of skill.

The tools available to the amateur generally are simple. A needle in a handle can do wonders if patiently and carefully used. A small chisel, usually ¼ inch and preferably a sculptor's chisel, is a valuable help for the cruder parts of the work. A soft paintbrush is useful for sweeping off the surface to keep a clear field. These are the basic essentials. Additional tools from the home workshop usually are more dangerous than useful, at least for the specimen. Even the lowly toothbrush probably has done more harm than good. This and even stiffer brushes should be used only with a light hand. Rotary tools normally should be avoided because even a small grinding wheel cannot tell rock from bone, while it covers the surface with dust and prevents the preparator from making the distinction.

With the simple tools outlined above, almost any preparing job can be tackled in time. Paleontological museums must try to save time in preparations work, and thus will have an assortment of power tools available. These tools particularly include small reciprocating drills, really miniature jackhammers. Some are driven by compressed air, others electrically. The air-driven types have two advantages: they tend to be more reliable and rugged, and the air blast from their tips keeps the work area clear of dust and chips. Several sizes are available, but all require the added expense of a compressor. The electric models are even smaller and most useful for extremely delicate work. Another important tool, a fairly recent addition to the preparators' arsenal, is basically a miniature sandblasting machine. These devices have done amazing jobs, some of which could not be handled by conventional tools. Hand tools are still the professional preparator's mainstay, particularly in the most delicate cases.

There are even more advanced techniques available. X-rays have been used to study specimens. Some see lasers as a new hope for a powerful, but accurate, preparing tool, though this device is in the future. In some cases, specimens may be destroyed to get a record of their shape. All of these are beyond the means of even the most advanced amateur and, indeed, out of the reach of many professionals.

Chemicals are a touchy subject in preparations work. Every preparator has dreamed of a magic fluid to destroy rock without harming bone, but it has not been found. The amateur should attempt nothing stronger than vinegar, keeping an eye open for possible bone damage even from this weak acid. It can be a valuable help on some specimens.

For the amateur preparator, we can only say: patience, care, and still more patience. It will pay off in the end.

Curation; or, the care and feeding of dinosaurs

The most important part of the work of a vertebrate paleontologist is the care of specimens, curation. The work is so important that "curator" is usually the job title. The task is twofold. First, the curator must make sure that nothing happens to the specimen. Second, and even more important, the information associated with the specimen must be preserved and indelibly tied to that specimen and only that specimen.

One type of information is important above all else: *exactly where was the specimen found?* It is essential to record an exact geographic location and the exact rock unit in which the fossil was found. Other information normally is added to this bare minimum: who collected it, when, and exactly what it is. Further notes will be added if the specimen is mentioned in a publication, especially if it has been "figured" or illustrated. These notes are particularly important if a new species designation has been based on this specimen. All the information is recorded on a numbered label. The number also is attached to the specimen in some way so it can be identified if separated from the label (such accidents are not rare). For serious museum work, the same information under the same number will be entered in a catalog, either a book or a card file. This additional record allows the label to be remade if it is lost or destroyed (again, not rare).

In most cases, *this information is far more important than the specimen itself,* and a specimen with its information (data) is far more likely to be important than one whose origin is unknown. A specimen with its data can be used in many ways. It may provide information about an entirely new animal, or about unknown parts of a known form. It may record the presence of an animal in a previously unsuspected geologic time interval and locality. Together with other specimens from the same locality, it can become a part of future studies. The relationship of one specimen to other animals from the site can yield valuable knowledge about the exact conditions under which they lived. If other specimens of the same species are found there, the specimen can form part of a study of variation in that species, a much-needed type of work.

A specimen without its data is useless for most work. Only if it provides evidence of a startling new animal or surprising new information about a known species is it even worth keeping. The data on a specimen are well worth the slight extra effort required. Curation is the difference between a collection and a pile of old bones.

How can you improve the curation of your collection? The first step is the most important. Decide that you are a collector, not an accumulator of curios. The rest will follow surprisingly easily.

Curation begins as soon as the specimen is picked up. It should be marked in some way so that you will know later where it was found. This precaution is especially important if collecting at several sites on a single trip. Large specimens may be marked with a felt-tip pen; many small specimens may be put in a marked container, as long as they are from the same locality.

Curation should be completed at home, while your memory is still fresh. Each specimen or group of specimens should be numbered individually. On large specimens a patch of white enamel can bear the number in permanent black carbon ink. If you expect to handle the specimen frequently, the number should be protected with a coat of clear nail polish. Small specimens may be placed in a box bearing their numbers. This method of numbering has the disadvantage of leaving no permanent identification on the specimen itself. A very important small specimen may be treated differently. Take a small vial with a cork that fits it. In the small end of the cork, stick an insect pin (common pins will do, but are not so suitable). Cut off the pin ¼ to ½ inch from the cork, and touch it to a small drop of rubber cement. Now touch the pin to the specimen in some place you will not need to examine closely. The specimen should stick to the rubber cement by surface tension. Arrange the specimen in a position that will allow easy study and let the cement dry. The cork can then be numbered, and inserted in the vial. This technique protects the specimen from dust and loss and permits direct attachment of the specimen number. Rubber cement is to be preferred for two reasons: it can be removed, if needed, by its thinner; and it is flexible enough to give some protection if the specimen hits the glass when it is pulled from the vial. Keeping the stub of the pin short helps to prevent this accident. The cork is also a handy stand and handle for looking at the specimen under magnification.

While the specimen is being numbered, write the label. The time required for the paint spot to dry is a handy period for this chore. Any paper and writing implement will do in a pinch, but museum people prefer paper of high rag content that will not crumble over the years or fall apart when wet. They always prefer permanent black carbon ink for two reasons. It is highly legible and will not fade. Pencil is preferable to washable inks or ball-point inks. You also may wish to safeguard your data further with a catalog, which may be either a blank book or a card file. The latter is handier, but the former is easier to store and individual cards cannot be lost from it. Some museums use both. The catalog entry should contain at least the same data as the label, and may contain them in fuller form, as more writing space is available.

The job of labeling and cataloging is made easier if you have pre-printed labels. These eliminate needless writing and serve as a guide to ensure including all important information. A personalized label also lends the collection a finished, professional appearance. There are many label forms in use. (See fig. 2 for one example.) A local

GEOLOGICAL SURVEY OF ALABAMA

Name _____

Formation _____

Horizon _____

Location _____

Collected by _____   Date _____

Type no. _____

Reference _____

FIG.   2.   Sample specimen label.

printer can provide labels (and catalog cards, if needed) at low cost; they are a worthwhile purchase.

It is important to recognize what constitutes an accurate description of locality. The ideal is simple: anyone on earth at any time in the foreseeable future should be able to find the locality precisely with no other information than the label and a good set of maps. This ideal is attainable but has been ignored all too often, even among earlier professionals. Such designations of localities as "Trego County, Kansas" or "Cretaceous of Texas" cause a never ending frustration for present workers. The first designation places the site somewhere within nine hundred square miles or so and gives no indication of the layer from which it was collected. The latter is even worse: it places the specimen somewhere within sixty million years of earth history and one hundred thousand square miles. Even localities that sound precise have proved impossible to revisit.

There are several ways to specify localities accurately. One is by latitude and longitude, down to the nearest second of a degree. This specification pinpoints the locality to within 100 feet for anyone on earth, whether in the same county or in Outer Mongolia. It does require very careful and accurate map work with a degree of skill that takes time to acquire. This method has other advantages: the location will not be incomprehensible even many years in the future and is suitable to electronic data processing. More and more museums are using this method for cataloging and research.

More familiar, and equally accurate, is a locality by land grid system (township, range, and section). Specifying a locality to quarter-quarter-quarter section (to the nearest ten acres) is accurate enough for most purposes and even to section (square mile) is above the standards of much past work. A location of this type can be determined by map work or by survey. Talking to the landowner often can save time. After all, the land probably is legally described in this way. Both the above methods are highly recommended.

Somewhat less accurate, but much easier in practice, is to give the mileage (within tenths of a mile) along a specified road from a specified point. This method is accurate and simple for road-cut localities. You can also give a direction and distance from some *permanent* landmark. These can be measured on a map or obtained in the field by pacing off the distance and getting direction from a compass. Road junctions are favorite landmarks; permanent survey markers (bench marks) are even better. Houses and barns, even with the owner's name, should be avoided. In any location involving roads, it is best to include a date (at least the year). Roads change courses occasionally; this tells a future hunter that he will need to know where, say, U.S. 84 was in 1938.

Below are listed examples of proper and improper localities. These are not real localities. Good examples of permanent locality descriptions are:

Wise County, Texas. Lat. 35°27′34″ N., Long. 96°42′13″ W. In gullies in pasture. Specimens from 0–8 inch green sandy clay layer in middle part of Paluxy Formation. [Such a locality would meet a future worker's wildest dreams. Just the latitude and longitude locate the site within 100 feet, and the rest should let an investigator put a hand right on it.]

Monroe County, Alabama, SW¼SE¼ sec. 11, T. 5 N., R. 4 E.; in cave on W side of N flowing creek. Specimens in white Ocala Limestone in cave roof. [Although the locality is originally specified only to within 40 acres, the additional information will help to pinpoint it.]

In road cut on U.S. Hwy. 84, 8.3 miles W of junction in Grove Hill, Clarke County, Alabama. Specimens weathered out of white limestone at top of cut. [Fairly good, and very easy to find.]

In ledge 27 feet above stream level on N side of Sipsey River 1500 feet upstream from mouth of Borden Creek, Bankhead National Forest, Winston County, Alabama. [Not the best but a very good substitute. The only worry is either a change in stream names or a change in water level. The measurement of 1500 feet is probably estimated by pacing.]

Here are some locality descriptions that look good on paper but are frustrating to find. Either they were not precise enough or have not stood the test of time. All are based on specimen labels the authors have encountered.

In field just NW of Cocoa Post Office, Choctaw County, Alabama. [Very accurate in its day. Unfortunately, the Cocoa Post Office no longer exists. Its location could be recovered by checking old residents or county records, but the problem is complicated because there were several old Cocoa Post Offices at different sites.]

7 miles NW of Grove Hill, Clarke County, Alabama. [Grove Hill covers roughly a square mile, so the area of search is at least that large. "NW" and "7 miles" both look like "eyeball estimates" and must be taken with a large grain of salt.]

700 feet SE of the old house on the Miller Place, near Greensboro, Hale County, Alabama. [Probably once very accurate but now more likely the most frustrating in the lot. There are probably fifteen families near Greensboro named Miller, most of whom have moved at least once. The Miller in question most likely moved "some place out west" thirty years ago.]

With all our complaints about inadequate localities, we must admit one thing: *Any locality information is better than none at all*. If there is any doubt in your mind as to the locality of the specimen, please note this on your label. Maybe you traded for it with someone whose veracity you don't quite trust. Maybe you found it in your pocket some time after a collecting trip and can't quite recall where it came from. *The only thing worse than no locality is a wrong locality*. An inaccuracy can mislead many future workers.

The study of the Eocene sharks of Alabama has long been confused by mislabeled specimens. These specimens were in private collections later given to the British Museum (Natural History). Woodward (1889D) accepted these erroneous labels at their face value and recorded several species from Alabama that do not occur in the state. White (1956) has done much to remedy this confusion, but some still exists. For example, Woodward's (1889D, p. 418) record of the giant shark *Carcharodon megalodon* from Alabama is based on a specimen from the Mediterranean island of Malta, despite the "Alabama" on its designation label (White, 1956).

Housing a private collection can be simple or elaborate, depending on ambitions and means. For a small collection, even a few cigar boxes are adequate. If specimens are small, inexpensive parts cabinets with many plastic drawers are particularly useful, although they will require occasional checking. For more elaborate housing, it is best to consult with professional personnel at a museum.

The most desirable features in any housing for fossils is that it be out of the reach of moisture and reasonably dust-tight and insect-proof. Insects do not eat fossils, true, but they do eat labels. A few scattered crystals of paradichlorobenzine ("moth crystals"), renewed as needed, will discourage their attentions.

A fossil collection does not require much care, but it should be checked periodically. This check will disclose any insect activity or moisture damage and will give early warning of the ravages of "pyrites disease." It may come as a surprise that there is a disease of fossils. Pyrites disease is not an infectious disease but more resembles cancer in its origin and effects. It is caused by the weathering of pyrite or related minerals, common in fossils as cavity fillings or as replacements of the bone. Exposed to air and moisture, pyrite oxidizes, producing minute quantities of sulfuric acid. This acid reacts with calcium carbonate in the bone to produce calcium sulfate (gypsum). The gypsum crystallizes in cracks and cavities and can quickly result in the complete shattering of the fossil.

Pyrites disease should be suspected if a formerly sound specimen begins to crumble or develops a white, powdery "bloom" on its surface. Treatment involves completely driving out moisture and then sealing the surface to prevent further entrance of moisture or air. The specimen can be baked for several hours at about 350°F., then completely sealed with shellac, starting with a very thin, penetrating mixture and working up to a thick surface coat. In a few "patients" the disease will have gone too far already or will refuse to be checked, but most can be saved by early and thorough treatment.

A properly curated collection can serve two purposes for you. First, it will impress acquaintances, particularly any visiting professionals. The latter will be far more willing to share knowledge with you, for you will have convinced them that you are willing to do careful work. Second, a well-cared-for collection can be a far more enduring monument to your memory than any inscription on stone. Much of our knowledge of the Eocene sharks of Alabama is based on three private collections that were donated or willed to the British Museum (Natural History) in the nineteenth century. One of these was the collection of J. W. Mallet, who was professor of chemistry at the University of Alabama and chemist of the Geological Survey of Alabama in the 1850s. The other two collectors, Enniskillen and Grey-Egerton, were

Britishers who acquired their collections by trade or purchase. All three of these men were exceptional amateur collectors, and the curation of their collections (though sometimes inaccurate, as noted above) was fully up to the museum standards of their day. Their collections have kept their memory alive far longer than have other accomplishments. Vertebrate paleontology in Alabama owes these three amateurs a considerable debt.

Identification; or, what it's all for

The specimen already has received a large amount of work. It has been found, collected, prepared, and curated. Now all this effort is about to bear fruit. You can begin to answer the question you started with: what is it?

It is difficult for professionals to write a section on identifying a fossil vertebrate. We must look back at all the years of training and experience and think of hours of frustrating search—a search that leads into dusty museum drawers that haven't been opened in years—a search that goes into musty library corners for books and articles rarely used. We remember the innumerable times when all research seemed in vain, when the specimen defied the best efforts, when it returned to its place still unknown. Years later, someone else, armed with better knowledge, may succeed, but for the moment the answer evades us.

Two thoughts spur us on. We can remember those times of pure joy when the veil of the unknown was nudged back another inch—when we have added to knowledge. The other is more sober—if we cannot explain clearly and simply what we have been doing all these years, we don't know ourselves.

The first fact to recognize is that not every fragment of fossil bone can be identified. The popular picture of the paleontologist's taking up a bit of bone and suddenly visualizing an entire unknown animal is a myth. It is a pretty myth that we wish were true, but we are not prophets. We are quiet, slow plodders, advancing step by minute step.

What is an identifiable bone? A single fragment out of the middle of a larger bone probably will hold its secrets forever. But give that fragment an edge or an end, a place where it meets another bone, and it may be identifiable. The probability of identification increases steadily the more complete the bone becomes. How fast the probability grows depends on what bone of the body it turns out to be. Ribs are poor prospects; vertebrae in some cases are little better. A complete limb bone is a pretty good candidate, as is any recognizable piece of skull. Teeth are the best of single bones; put them in series in a jaw, and success is almost certain. From this point, more bone will add still more details and more precision to the final identification.

As the last statement suggests, not all identifications are equally precise. Some specimens can be placed in a particular species; others give less, but still useful, information. Later parts of this work will add several records of animals previously unknown from Alabama. Most of these are based on very fragmentary specimens that cannot be identified precisely, yet they give valuable new information about the life of Alabama's past. Such specimens should not be scorned.

Two questions normally are considered in the identification of a fossil vertebrate. Their answers are not necessarily separate but generally are. The first is: *What bone in the body is this?* The second: *What animal has this bone in precisely this form?*

"What bone in the body is this?" may be a simple or a difficult question to answer. If it can be answered, half the battle is won. Many animals can be quickly eliminated; they may not even have such a bone, or they may have it in a minute, degenerate form that is no longer useful. If you know what bone it is, you will also know exactly what function it served in the body. A little study will indicate just how the bone served that function, so you learn something of the habits of the animal. From this point you have an excellent chance to extend the identification to a more precise point.

An example from the experience of one of the authors might be instructive. He picked up a bone in a road cut in sediments of Pleistocene age (the time of great ice sheets in the north). The bone was not complete, but one end and about two-thirds of the shaft were preserved. The shape of the end *indicated* that it was the proximal end (end nearest the body) of a humerus (upper bone of the forelimb). Other characters of the end *indicated* that it belonged to a mammal, and its size *indicated* something roughly the size of a domestic cat. The shaft was peculiar in two ways. About one-third of the way from the end, it was sharply bent, *indicating* a bowlegged stance for the animal. The shaft was flattened, so that if placed back in the body, it would have been much wider vertically than horizontally. The significance of this feature was not realized until much later. Had it been realized, the task would have been much easier.

Still, only certain cat-sized animals are also bowlegged. Armadillo immediately came to mind (by the way, the specimen came from Texas). An armadillo skeleton was quickly resurrected from the museum collection, and its humerus was directly compared with the unknown bone (this kind of comparison is something of a court of final appeal in paleontology). Nothing doing. They weren't even remotely alike. Badger? No skeleton was in the museum collection. The paleontologist cooled his heels until he visited another museum with a more complete skeleton collection. A badger was quickly located but failed the comparative test also. It was time for a reconsideration of the facts.

What about that flattening of the shaft? That is a common feature in the limbs of aquatic animals; it makes the limb a more effective oar. An otter skeleton might give some information. Without much hope, he pulled one out. The bones matched perfectly, even in size. Another specimen had yielded its secrets; another piece of the puzzle was added.

There are no magic secrets in identification. One comment made earlier needs some clarification. We indicated that teeth are the most identifiable part of a vertebrate fossil. Why should this be so? The teeth of an animal permit it to acquire and process the most important part of its environment—its food. For this reason teeth closely reflect the habits of the animal: flat grinding teeth for plants, piercing and cutting teeth for meat. Other types of teeth may show even more specialized diets, such as carrion or shellfish.

In general, teeth with several points (cusps) in a complex pattern belong to mammals; with the information on diet that the tooth itself provides, you are already far toward identification. A simple tooth with a single point may belong to a fish, an amphibian, or a reptile (perhaps even a bird, but this would be an extremely rare find). Identification of teeth may be easy or complex. See the descriptions for further information. A large part of it is memory and observation. *The more bones you have studied, the quicker and easier identification becomes.*

Where does this leave the amateur, who probably has not seen many bones except in chicken or steak? Without a vast collection of already identified bones to work from, the amateur is behind the professional, a long way behind, and can catch up only by devoting full time to study. However, most amateurs study fossils just during spare time.

Although difficult, the task is not impossible. Step by step, you can learn an amazing amount. Start simply. Collect specimens. This book will serve as a starting point in identifying them. Put four or five on the table in front of you. Study them carefully, trying to make a mental image of each detail. Then thumb through the figures in this book. Does anything look familiar? If it does, look more carefully at the figure and the specimen. Are they alike in every detail or only slightly different? Read the description of the species named in the matching picture. It will supply details that cannot be shown in the best picture. Does the specimen fit the description as well as the picture? You are on the right track. Your identification may turn out to be wrong later, but you have made a significant beginning. You are pushing back the unknown. Every paleontologist has started in some fashion like this; he once had absolutely no idea what he was doing. The first few identifications probably would look ridiculous now; they certainly looked ridiculous to his professors! These mentors corrected his mistakes with patience and encouragement; you must correct your own or find someone who can do it for you.

Where can you find help to check your identifications or to set you on the right track when stumped? The Geological Survey of Alabama is probably the most likely place. Even though someone there may not be able to help you directly, the Survey maintains close contact with all persons in Alabama who are working in geological fields, including vertebrate paleontology. In most cases, however, the Survey can answer your questions.

Until very recently, help in vertebrate paleontology was not easy to find in Alabama. At this time (1980), work on vertebrate fossils is going on at The University of Alabama (where the Geological Survey of Alabama also is located), Auburn University, and the University of South Alabama. It is impossible to predict where work will be carried out in the future. Vertebrate paleontologists are beginning to realize more and more the potential of Alabama. With the help of the many amateur paleontologists in this state, our knowledge of the life of the past will expand rapidly.

Paleoecology; or, where do we go from here?

Earlier we indicated that one of the major uses of paleontology was the study of past environmental conditions, or paleoecology. Therefore, paleoecology is one of the objectives we must achieve.

Paleoecology is a rapidly growing branch of science. Not only paleontologists take part, but students of sedimentary rocks play a considerable role. Many techniques have been used; far too many to list here, but most of these are based on one common assumption: *Anything found in the earth is to be explained in terms of biologic and physical processes that can be observed working now.* Some find a briefer statement more satisfactory: *The present is the key to the past.* This assumption perhaps can be described best in a series of questions that the paleoecologist must answer.

What is known about this ancient environment? On the basis of what is known now, what modern environment seems to be a counterpart? What information does this modern environment provide regarding the ancient one? In particular, what should we find in the ancient one that we haven't found?

Can these "missing" characteristics be found in the environment being studied? If so, maybe the right track is being followed. If not, why? Is the analogue incorrect and a different modern environment needed for comparison? Or are we able to provide valid reasons why we can't find what is expected?

If the science of paleoecology can become an exact science, so that we can understand the forces that have modified the environments of the past into those of the present, we will be able to change that statement further. We are on the brink of saying *the past is the key to the future.*

Vertebrate paleoecology is a science in its infancy. Although a few scientists had been interested in it for over a century, most vertebrate paleontologists seemed to have felt that enough work was entailed just identifying collections, especially considering all the time, study, and effort needed in securing collections and preparing them for study. Now a growing feeling in the profession is that enough information is available to begin to put it to work in reconstructing ancient environments.

Since 1950 a large number of classic studies in vertebrate paleoecology has been made. The number is so large that it has become impossible to cite them all. We can do no more than give some examples, more to demonstrate the highly varied methods that have been applied than to show the conclusions reached.

One worker (Lundelius, 1969) had a large number of specimens of an extinct llama from a single locality in southern Texas. Careful measurements and study of tooth wear in the individual specimens led to a surprising conclusion; it was possible to estimate, within perhaps a month or two, the age of the individual animals at death. The locality seemed to represent a prehistoric waterhole. If the llamas were in the area throughout the year, we would expect that occasionally one would become mired in the waterhole or be killed by predators at its brink. With a large number of specimens, one llama of every conceivable age should have been represented.

This was not the case. Rather, all the specimens tended to fall into very narrow age groups. No newborn specimens were found, but weanlings were common. Yearlings were not present, but animals a few months over a year old were. For older animals, distinctions were not so clear-cut, but there were groups of specimens a little over two, three, and even four years old.

How could such a distribution be explained? Two elements seemed necessary: a sharply defined breeding season for these ancient llamas, and yearly migrations that would have placed them near the waterhole for only a short time each year. Such migrations presumably would have been caused by the seasons, yet other evidence indicated that summers were fairly cool and winters quite mild at this time.

Another clue was to be found in the bones from this ancient waterhole; there were fossil fishes whose living relatives exist only in permanent bodies of water. The suggestion was made that this locality represented one of the few waterholes in the area that did not dry up in the dry season. Animals would then congregate about it at the height of the dry season when they were in their weakest condition. Such circumstances could also explain the large amounts of bone at this site—weakened animals are least able to free themselves once mired, and many would have died in any case.

From this pattern an important conclusion emerged. While summer and winter had little meaning in this area at this time, wet and dry seasons were pronounced. The environment might not have been unlike that of the Serengeti Plains of Africa, renowned today for wildlife. As a result of this type of investigation, another fragment had been added to our knowledge of the past.

These remains dated from the Pleistocene Epoch, the very recent time of great ice sheets in higher latitudes. The study of fossil vertebrates has led to some surprising conclusions about the climate of those times. For many years students of the Pleistocene assumed that the great glaciers brought the conditions of the far north with them. A belt of tundra, frozen treeless plains, extends for over 300 miles in front of existing ice sheets of similar size. Should tundra not also have extended in front of the Pleistocene glaciers? This would have made climatic conditions in northern Alabama very much like those in the Canadian Arctic only twenty thousand years ago.

A growing body of evidence indicates that this picture is not true. The first inklings of its falsity were due to the work of Hibbard (1944B). Among other things, he found the remains of giant land tortoises in southwestern Kansas, within 200 miles of maximum glacial advance, in beds that matched the very time of that greatest ice advance. Tortoises today cannot survive if their body temperature falls below 40°F.; there is no reason to think their ancestors were more resistant. Small land turtles can survive cold winters by burrowing out of reach of the frost. This option is simply not available to their larger cousins, some of whom had shells over six feet long. The only possible conclusion seems to be that although nearly half this continent was buried under thousands of feet of moving ice, the climate beyond that ice featured milder winters than those of today.

This idea is dangerously close to controversy; it is time for a note of caution. The conclusion reached above is the subject of considerable debate. It is not accepted by all vertebrate paleontologists. Even less is it accepted by some workers in other fields. Many students of ancient plants still believe that the older concept of considerable cooling in front of the glacial ice has merit and base their case on much solid evidence.

Such disagreement is common in science. Many workers are seeking their own ways toward knowledge, and each path is different. New ideas are tried in the fires of controversy, and many fail. Others emerge tempered into new instruments in the search for truth. Often they are changed by their ordeal. Someday, someone will weld the two ideas noted above in a way that will probably surprise most of us but will fit all the available information into a single picture. Part of the answer almost certainly lies in Alabama.

Systematic Paleontology

The world of taxonomy and nomenclature (scientific classification and naming) can be a strange and confusing one to the amateur. Yet its objectives are clear, simple, and necessary. Nomenclature (literally, "name calling") provides each kind (species) of animal with a name that belongs to that kind alone. Taxonomy (classification) places each of these species in a scheme that expresses its relationship to every other animal. The same principles apply to all organisms, but we limit our treatment herein to animals. The accomplishment of these aims to the best of our ability is essential to an understanding of the life of the past.

The basis of the whole system is the species, especially when dealing with living animals. Just what constitutes a species is the subject of considerable debate. All workers, we believe, would agree with the following definition. *A species is a kind of animal that is distinct from all other kinds of animals.* Debate on the "species concept" tends to center on two points. Just how *distinct* must a species be, and how do they get distinct and stay distinct? As long as there are two taxonomists left alive to argue, these two points probably will continue to be debated.

Closely related species are placed together in a *genus.* The name of the genus (generic name) and the name of the species (specific or trivial name) together make up the proper scientific name of the species. For example, the domestic dog is named *Canis familiaris.* The name *Canis* belongs to the genus; it indicates that this animal is very close kin to the wolves, coyotes, and jackals, who share certain common characteristics. The trivial name *familiaris* specifies still further; this is the domestic dog.

The format in which these names are presented is invariable. Both are in *italics* when set in type; in manuscript, they are underlined. The generic name comes first and always begins with a capital letter; the specific name is second and is never capitalized (except for species named for persons in some older literature).

In careful work, the name should be expanded further. Our example would be stated as *Canis familiaris* Linnaeus, 1758. The additions include the name of the describer of the species (the one who first considered this a distinct species and named it in this particular format) and the date when the describer's name and the description of the species were published. Our example happens to be credited to the man who first established the system we now use and comes from the book in which the system was established.

Related genera are placed in the same family; related families in orders; and so on through classes and phyla (singular, phylum). All the animals discussed in this book belong to the Phylum Chordata.

Since the establishment of this system in 1758, our knowledge has expanded greatly. We know much more about the relationships of animals than Linnaeus did, so expansion of his system has been both useful and necessary. In almost all cases, this expansion has been done by placing modifying prefixes before the terms Linnaeus used. There are three of these prefixes in common use; in descending order of size, they are *super-*, *sub-*, and *infra-*. Thus a superorder would be a grouping of closely related orders, but on a smaller scale than any grouping involving the term "class." A suborder would be a division of an order, while an infraorder would be a division of a suborder larger than a superfamily.

Linnaeus' scheme of classification has held up remarkably well for more than two centuries, with only minor modifications. Some terms have been added, but these are usually the subjects of debate about their precise significance. At present two added terms are used in vertebrate paleontology to some extent: the *cohort*, ranking between order and class; and the *tribe*, between genus and family. Neither will be used in the classification adopted here, as these superfine divisions are not needed for our purposes. They are included only because they might be encountered in further reading.

Why is all this necessary? Why not call a dog a dog and have done with it? There are two essential justifications. First, we hope to establish, and have largely achieved, an international scientific language. The English word "dog" might be incomprehensible to a worker in Ulan Bator or Timbuktu; *Canis familiaris* would be understandable regardless of the taxonomist's mother tongue. Second, many animals lack popular names. This is true of animals from remote regions and still more true of fossil animals. It is also true of many Alabama animals, such as the many species of snails native to the state.

The system outlined is magnificently simple in its concept. Its application has been neither simple nor, in some cases, magnificent. The vast amount of knowledge that has accumulated about the animals of the world, past and present, is beyond the ability of any single worker to comprehend. For this reason there has been overlap; two different workers may propose different names for the same animal or use the same name for different animals. In such cases, confusion results. This confusion has tended to accumulate over the centuries and has placed the entire taxonomic system in danger of collapse. Alternative systems have been proposed from time to time almost from the date of Linnaeus' first concept of the order of nature. None yet has proved acceptable to even a minority of students. Perhaps an alternative system will be adopted someday. Its proposer will have to be a genius of greater rank than Linnaeus and not only will have to devise the system but also will have to be a persuasive writer to propound his ideas and a diplomat to influence even a minority of colleagues to abandon the habits of a lifetime. This event seems unlikely. Until it occurs, and for long after, we are left with a strong need to preserve the Linnaean system.

Still more confusion in the naming of animals has resulted from the debate over the exact nature of a species. Some workers have tended to "split" species, naming every minor variation as a new form. Others have "lumped" species together into conglomerations of quite diverse animals. The greatest vogue of "splitterism" occurred in roughly the first four decades of this century. In vertebrate paleontology at this time, " 'n. sp.' [new species] means merely 'new specimen,' " as Welles (1962), translating another author, has aptly remarked. We are only now emerging from the clutter of this era.

In cases of multiple names for the same animal, how do we decide which name is to be applied? This decision is made by strict application of the *law of priority*. The first name applied to an animal is the valid name, unless it is unusable for reasons noted below. Application of this law has led to confusion in some cases. Names long in use and firmly entrenched in the literature have been found to be junior synonyms of very obscure names. For example, the tiny Eocene horse is known almost universally as *"Eohippus."* This name is now known to be a junior synonym of *Hyracotherium,* a genus based on teeth from Europe.

This rule applies to the same name used for different animals. The first use of that name will stand, and all other uses of the name for other forms are invalid. A common shark in the Alabama Cretaceous was long known as *Corax* until that name was found to have been applied previously to European ravens. Nevermore would *Corax* be a valid name for a shark, and this genus was renamed *Squalicorax.* This case also affords an example of the law of priority as applied to synonyms. Two workers independently realized that *Corax* as applied to a shark was a preoccupied (previously used) name, and each renamed the genus, one as *Squalicorax* and the other as *Anacorax.* Because *Squalicorax* was published a few months earlier than *Anacorax,* it is the valid name for this genus.

Application of these rules can be extremely complex. To add further to the complexity, the International Commission on Zoological

Nomenclature is empowered to suspend the rules in cases where it is clear that their application would create more confusion than permitting an exception to exist. In general, nomenclatural problems are the bane of professional workers. It often seems that we spend one-fourth of our time determining what the specimen is, and three-fourths deciding what name should be applied to it. Solving such problems requires diligent, careful, and thorough research in the literature of zoology, often in obscure places. It is probably best left to the professional. In most cases, the amateur would be advised to adopt some particular set of nomenclature as a model and stick to it.

Vertebrate paleontology has another severe problem with the scientific naming of animals—the indeterminate name. During the last half of the nineteenth century, the study of fossil vertebrates passed through a phase when every wretched fragment of bone seemed to receive a name if it were at all different from what had been found before. Often it has been impossible to decide exactly what animal is represented by these inadequate specimens. The problem becomes even more severe when the types (original specimens) have been lost, and decisions must be made on sketchy descriptions and poor figures. Here lies the main reason for the careful preservation of any specimens that have been described in the literature—the specimens themselves contain far more information than any description or illustration can convey.

In any event, we are left with a mass of names of uncertain application. Some of them may be senior synonyms of much better known names based on adequate material, but present methods cannot determine their status. How are we to deal with such names? The best procedure, professionally, is to ignore them once they are known to be indeterminate, unless some future worker, with better methods, can find exactly what animal is represented. Unfortunately, this makes for much instability in the nomenclature. Under the present rules, the only means to rid ourselves of such useless names is not to use them for a considerable period of time. They then become "forgotten names" (officially *nomina oblita*) and cannot be resurrected.

The nomenclature used in this classification will not satisfy all professionals. Some points have been debated for years and will be for years to come. We have attempted to follow the most recent classifications available for each group and to cite the names that are in current use.

Fossil Vertebrates
of Alabama

PHYLUM CHORDATA

This vast group of animals includes all forms that have, at least at some stage of development, the following three characteristics: gill slits, a stiffening rod (notochord) in the body, and a hollow, single nerve cord located dorsally to the notochord. The Phylum Chordata includes several primitive or degenerate groups in addition to the vertebrates.

SUBPHYLUM VERTEBRATA

In addition to the basic chordate features, the vertebrates are distinguished by the development of segmented stiffening structures (vertebrae) around the notochord. These may be composed of cartilage or bone and may completely replace the notochord. The nerve cord has an anterior expansion, the brain, partly or completely enclosed in a cartilaginous or bony braincase (or cranium). Their geologic range is at least Middle Ordovician to Recent. If the doubtful conodonts are included, they are first found as fossils in the Upper Cambrian.

Class Agnatha

Primitive jawless vertebrates, the Agnatha are known from the Middle Ordovician through the Recent. The only living representatives of this class, the mostly parasitic lampreys and hagfishes, have no bony parts and are known as fossils from the Pennsylvanian. The jawless fishes with bony skeletons are confined to the Paleozoic and are most common in the Silurian and the Devonian. No fossil Agnatha are known from Alabama.

Class Placodermi

The placoderms are very primitive jawed fishes, distinguished from later groups by technical details of jaw structure. They range from the Silurian to the Permian and have no living representatives. The placoderms are unknown in the fossil record of Alabama.

Class Chondrichthyes
cartilaginous fishes

As their popular name implies, the Chondrichthyes have no true bone in the skeleton. The most common hard parts are teeth, isolated occurrences of which constitute most of their fossil records. Scales and fin spines, consisting of bony material, also are found. The remainder of the skeleton is composed of cartilage (gristle). In a few cases, this tough, resilient material is reinforced by calcium deposits, especially in the vertebrae and jaws. Such calcified cartilage is easily distinguished from true bone, especially in microscopic study of thin sections. It consists of thin, concentric layers and never shows the contorted tubules of true bone. Cartilaginous fishes are known from the Devonian to the Recent. Their fossil record in Alabama is extensive and includes the oldest of the known Alabama vertebrates.

Subclass Elasmobranchii
sharks, skates, rays, and sawfishes

These are the "normal" cartilaginous fishes and include all of the Chondrichthyes except for a few aberrant, highly specialized forms of uncertain derivation. They are known from the Devonian to the Recent.

Order Cladoselachii
primitive sharks

The cladoselachians include the earliest known sharks. Remarkable specimens from the Devonian of Ohio preserve not only the teeth but the entire body, with traces of most of the internal organs. They are distinguished from other sharks by the very primitive nature of the jaw structure, as in the specimens from Ohio. Identifications of most known cladoselachians are based on teeth alone. The range of the order is Devonian to Permian.

FAMILY CLADOSELACHIDAE

Most of the known cladoselachians are in this family. *Cladoselache*, the typical form, is known from complete specimens from the Devonian of Ohio. Isolated teeth usually are placed in the genus *Cladodus*, which serves for a multitude of poorly known, probably only remotely related forms, some of which seem to pertain to the Ctenacanthidae.

"Cladodus" magnificus Tuomey (fig. 3)

C(ladodus) magnificus Tuomey, 1858A, pp. 39–40, figs. C, Ca, Cb. Name in figure captions.
Cladodus magnificus, Newberry and Worthen, 1866A, p. 24, pl. 1, figs. 6, 6a; Newberry, 1889A, p. 216.
non *Cladodus? magnificus* Claypole, 1894A, pp. 137–140, pl. 5, (=*Cladodus claypolei* Hay, 1899E, p. 137).

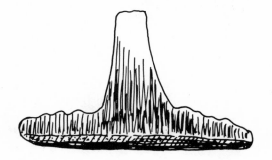

FIG. 3. *"Cladodus" magnificus* Tuomey. From Lauderdale County, Alabama. After Tuomey, 1858A. *c.* x1.

This is a large species, based on a tooth over an inch long collected by Tuomey in Lauderdale County, Alabama, from rocks near the contact between the Chattanooga Shale and the Fort Payne Chert. The age may be either latest Devonian or earliest Mississippian, probably the latter (Drahovzal, oral communication). The type specimen was lost, probably when Federal forces burned The University of Alabama in 1865, and we have only the figures by Tuomey (1858A) and Newberry and Worthen (1866A) for reference. The latter specimen presumably is in the collections of the Illinois Geological Survey but has not been examined by the authors.

One drawing by Tuomey (1858A) is reproduced in fig. 3. The crown has a tall, fluted main cusp with a pronounced shelf on one side, presumably the inner (lingual) surface. Tuomey's figures do not agree on the presence or absence of accessory cusps. Those of Newberry and Worthen (1866A) show strong accessory cusps.

References of this species to *Cladoselache* are based on confusion of this species with the form described by Claypole (1894) from the Devonian of Ohio. In general, Tuomey's work has been ignored by later students.

"Cladodus" newmani Tuomey (fig. 4)

Cladodus Newmani Tuomey, 1858A, p. 39, fig. B.
Cladodus newmani, Newberry, 1889A, p. 216, "too imperfect for identification."

FIG.   4.   *"Cladodus" newmani* Tuomey. From limestone of Mississippian age near Huntsville, Alabama. After Tuomey, 1858A. *c.* x1.

Smaller than *"Cladodus" magnificus,* this species has very weak accessory cusps. The specimen was found by Tuomey (1850B) in "a fine collection of the fossils of the Carboniferous rocks about Huntsville . . . in Dr. Newman's cabinet." It would presumably be of Mississippian age. There is some doubt whether Tuomey ever added the specimen to his collection, as it may have remained in Newman's. In either case, the tooth is presumed to have been lost. Tuomey's (1858A) illustration is reproduced in fig. 4.

FAMILY CTENACANTHIDAE

Cladoselachians with highly ornamented fin spines are referred to this family. They range from Devonian to Permian.

Ctenacanthus elegans Tuomey (fig. 5)

ctenacanthus [sic] *elegans* Tuomey, 1858A, p. 38, text-fig. A.
?*Ctenacanthus angustus* Newberry, 1889A, p. 81; Hussakof, 1908A, pp. 44–45, fig. 19.

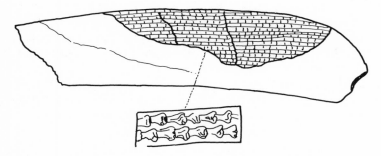

FIG. 5. *Ctenacanthus elegans* Tuomey. "Represents the fossil at natural size, with a portion of the enamel remaining." After Tuomey, 1858A.

The partial fin spine on which this species is based was not collected at the same locality as was the type of *"Cladodus" magnificus* Tuomey. The spine shows several rows of enamel beads, each bead roughly triangular in shape (fig. 5). The incomplete nature of the specimen did not permit Tuomey (1850A) to determine the number of rows. The ornamentation of this spine is very similar to that of *Ctenacanthus angustus* as figured by Hussakof (1908A) and may be a senior synonym of that species. The type specimen must be presumed to be another casualty of the War Between the States. Again, we reproduce Tuomey's figure, all that is known so far of the species. These spines may prove to come from the same animal that produced the teeth of *"Cladodus" magnificus*. *Ctenacanthus elegans* would then be the senior, and valid, name for the species.

Order Selachii
sharks

The sharks are important today as marine predators. The earliest known sharks, in the strict sense, are Late Devonian.

Suborder Hybodontoidea
hybodont sharks

The hybodonts are primitive true sharks. The teeth of hybodonts can be readily distinguished from those of more modern sharks by the presence of multiple openings for blood vessels and nerves in the roots. Modern sharks have only a single opening. Hybodonts range from Upper Devonian to Eocene.

FAMILY HYBODONTIDAE

The "normal" hybodonts, this family presumably contains the ancestors of all later sharks and rays. The dentition is extremely varied, ranging from powerful tearing and slashing teeth in carnivorous forms to flattened crushing plates in those feeding on shell-bearing animals. The hybodonts were the most prominent marine carnivores in the late Paleozoic and early Mesozoic. By Cretaceous times, they had become subordinate to their descendants, the galeoids, and survived only in specialized niches. There was a brief flowering of hybodonts in the Early Cretaceous (Patterson, 1967; Thurmond, 1969, 1972), confined largely to fresh waters. A few marine forms lasted until the close of the Cretaceous, and one genus, *Synechodus*, persisted into the Eocene.

Dicrenodus wortheni (Newberry and Worthen) (fig. 6)

Carchariopsis wortheni Newberry and Worthen, 1866A, p. 69, pl. 4, figs. 14, 14a.

Presumably Mississippian in age, this species is based on a single tooth from near Huntsville. The tooth rather strikingly resembles those of the much later and unrelated white sharks, *Carcharodon,* hence the name originally applied to the genus (*Carchariopsis* means "looking like *Carcharodon*"). The main cusp of the tooth is broad, triangular, and coarsely serrated if present (Blot, *in* Piveteau, 1966,

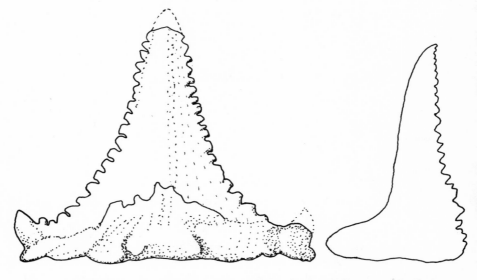

FIG.   6.   *Dicrenodus wortheni* (Newberry and Worthen). Tooth, inner and side (outline only) views. After Newberry and Worthen. x1.

p. 730). The use of the name *Dicrenodus* for this genus is prevalent now (1980), but we adopt it with some misgivings in view of the fact that *Carchariopsis* is the older name.

FAMILY PTYCHODONTIDAE

These unusual fishes, known largely from teeth, were long classified with the skates and rays (e.g., Berg, 1958). Because of the recent work of Casier (1947, 1953) they are now ranked as strangely modified hybodonts. The dentition is a broad pavement of roughly rectangular teeth ornamented with ridges and tubercules, highly suited for crushing shells.

Ptychodus mortoni Mantell (fig. 7)

Ptychodus mortoni Mantell, in Morton, S. G., 1842A, p. 215, pl. 11, figs. 7, 7a.
Ptychodus mortoni Ag[assiz]., Tuomey, 1858A, p. 261.
Hemiptychodus mortoni Jaekel, 1894A, p. 137; Hay, 1902A, p. 317.
Ptychodus mortoni, Leidy, 1868A, p. 206; Applegate, 1970, p. 393.

FIG. 7. *Ptychodus mortoni* Mantell. Photograph of specimen in collection of the Geological Survey of Alabama. x1.

This species is represented by numerous teeth from the Eutaw, Mooreville, and Prairie Bluff Formations of Alabama. The teeth usually are less than one inch across, have highly domed crowns, and show crude striae radiating from the center of the crown.

Ptychodus polygyrus Agassiz (fig. 8)

Ptychodus polygyrus Agassiz, 1835, p. 55; 1843B, (actually 1839), vol. 4, p. 156, pl. 25,
  figs. 4–11, pl. 25*b*, figs. 21–23.
?*Ptychodus* (sp. ?), Tuomey, 1858A, p. 261.
Ptychodus polygyrus, Applegate, 1970, p. 393, fig. 179A.

FIG. 8. *Ptychodus polygyrus* Agassiz. Photograph of specimen in collection of
Geological Survey of Alabama. x1.

A large species, *Ptychodus polygyrus* has teeth that are 2 inches
across in many cases and larger in some examples. The crown is much
less domed than that of *P. mortoni* and shows powerful ridges running
from side to side across the crown. The border of the crown is covered
with small tubercules, as in fig. 8.

Suborder Heterodontoidea
frilled and Port Jackson sharks

This suborder is composed of the living sharks *Chlamydoselache*
(frilled shark) and *Heterodontus* (Port Jackson shark) and their fossil
relatives. They are barely modified descendants of the hybodonts and
are lumped with the hybodonts in many classifications (e.g., Blot *in*
Piveteau, 1966). The frilled sharks are not common as fossils and are
unknown in Alabama rocks.

FAMILY HETERODONTIDAE

Heterodontus sp. cf. *H. woodwardi* Casier (fig. 9)

Heterodontus cf. *woodwardi* Casier, White, 1956, p. 128.

FIG. 9. *Heterodontus woodwardi* Casier. Lateral tooth. After Casier, 1946. *c*. x5.

White (1956) reports this shark from rocks of late Eocene age (Jackson) of Clarke County (?) and notes that the single tooth in the British Museum (Natural History) collections (P. 30514) is almost identical with one from Belgium figured by Woodward (1891A, pl. 3, fig. 1). As White did not figure his specimen, we copy the figure by Casier (1946, p. 45), who referred this figured tooth to his new species, *Heterodontus woodwardi*. The teeth of *Heterodontus* are always broad, domed, crushing plates. Many earlier authors call this genus *Cestracion*.

Suborder Galeoidea
true sharks

Most of the fishes commonly considered sharks are in this suborder. There are five gill slits on each side of the body and only a single nutritive canal in each tooth root. Almost all are purely carnivorous— the "feeding frenzy" of a group of sharks around a carcass is one of the most fearsome sights in modern seas. Modern adult sharks range in size from small, inoffensive creatures less than a foot in length to 18 feet of saw-toothed death in the white sharks *(Carcharodon)*. Even larger are the whale sharks, perhaps attaining 60 feet, but these feed on plankton.

The most common shark fossils are teeth. In some cases, calcified vertebrae will be found. These are roughly checkerlike discs (up to 4 inches across) with weakly fluted sides. The center will be nearly pierced through. Shark vertebrae are not particularly distinctive, although some can be identified (see Applegate, 1970). Fortunately, the more abundant teeth are very distinctive and readily identified.

The tiniest of shark fossils are the *dermal denticles*. Actually, these are scales and give shark skin its "sandpaper" feel. Most of them are less than 1/100 inch across and can be collected only by very specialized methods. They are extremely abundant, but very little work has been done with them.

The earliest true sharks are Middle Jurassic in age, and they are abundant and varied today. Their fossil record in Alabama is extensive in time, space, and numbers, so that shark teeth are the most familiar vertebrate fossils to most Alabama collectors. From a scientific standpoint, only the turtles and the basilosaurs have been more intensively studied in the state.

FAMILY ORECTOLOBIDAE
nurse and carpet sharks

These are very primitive and somewhat atypical true sharks. Several teeth of each row are functional at any one time, and each tooth is more of a rasping than a piercing or tearing weapon. The base of each tooth is much swollen and characterized by the presence of a strong ridge on the crown. This ridge bears a single large cusp and many small ones. The group feeds mainly on invertebrates. The nurse sharks are quite inoffensive and should not be confused with the dreaded "gray nurse" of Australia, which is actually a sand shark. The family ranges from Jurassic to Recent.

Ginglymostoma serra (Leidy) (fig. 10)

Acrodobatis serra Leidy, 1877A, p. 250, pl. 34, figs. 10–13.
Acrodobatis obliquum Leidy, 1877A, p. 250, pl. 34, fig. 14.
Ginglymostoma serra, Woodward, 1889D, p. 348, pl. 16, fig. 9.
Ginglymostoma obliquum, Leriche, 1942B, p. 52; White, 1956, p. 124.

FIG. 10.   Ginglymostoma serra (Leidy). After Leidy, 1877A. x2.

Woodward (1889D) reported this species from the Jackson (upper Eocene) of Clarke County. Leriche (1942B) and White (1956) both reidentified Woodward's two teeth as Ginglymostoma obliquum. These two species were thought to differ in the inclination of the main cusp, erect in G. serra and inclined in G. obliquum (hence the name). A series of teeth from the Gosport Sand at Little Stave Creek, Clarke County, shows a continuous gradation in main cusp inclination between the two types previously distinguished. Thus, they are considered synonyms here.

Teeth of Ginglymostoma, the nurse shark, are difficult to describe verbally. The base is greatly expanded, with a lobe on the buccal (outer) face. The main cusp has one or more accessory cusps on either side (five to nine in G. serra).

Ginglymostoma sp. cf. *G. blankenhorni* Stromer (fig. 11)

Ginglymostoma blankenhorni Stromer, 1905E, p. 166, pl. 15, figs. 28–31; Arambourg, 1935F, p. 422, pl. 20, figs. 5–6.

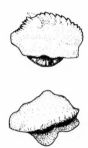

FIG. 11.   *Ginglymostoma* sp. cf. *G. blankenhorni* Stromer. Tooth, BSC 780, near Peterman, Monroe County, Alabama. Buccal (outer) and occlusal views. x3.

A new occurrence for Alabama, this species is based on teeth from the Tallahatta Formation, ¼ mile southeast of Peterman, Monroe County. *Ginglymostoma blankenhorni* has been reported only from the Eocene of North Africa (Stromer, 1905E; Arambourg, 1935F). The African teeth, and the Tallahatta specimens, show an inclination of the main cusp considerably greater than that of "*G. obliquum*" (=*G. serra*). The cusp immediately posterior to the main cusp is much larger than the other accessory cusps in the African material but not in the Alabama specimens.

These Alabama teeth may represent a new species but are referred provisionally to *G. blankenhorni* until more material is collected. The specimens are in the collection of the Geological Survey of Alabama.

FAMILY ODONTASPIDAE (CARCHARIIDAE)
sand and goblin sharks

These are primitive, but typical sharks. The body is slender and the tail unusually long and powerful. The anterior part of the head (rostrum) is long and pointed, grotesquely so in the living goblin shark, *Mitsukurina*.

The name of this family, and of some of the genera within it, has been in a state of confusion for many years and probably will be for some time to come. A detailed rehash of this long-standing controversy is not necessary here. Essentially, the points of debate center around the exact animals intended in certain very old descriptions. Paleontologists generally refer to the family as Odontaspidae while students of living sharks prefer Carchariidae. This latter name should not be confused with Carcharhinidae, a quite distinct and important family.

The teeth generally show a very slender and extremely sharp main cusp, often sigmoid (S-shaped) in side view. There is one pair of accessory cusps (two in a few cases). The root is deeply forked and has a very prominent shelf on the lingual (inner) side. This shelf is vertically divided by a strong groove. When observed from the lingual side, to quote a fine old Victorian paleontologist (Cope, 1875), the view is "not unlike the pygal region of a Hottentot Venus." We will not elaborate further; a good physical anthropology text will provide details.

Scapanorhynchus raphiodon (Agassiz) (fig. 12)

Lamna (Odontaspis) raphiodon Agassiz, 1843B, vol. 3, p. 296, pl. 37a, figs. 12–16.
Lamna elegans, Tuomey, 1858A, p. 261 (not of Agassiz, 1843B).
Scapanorhynchus texanus, White, 1956, p. 124.
Scapanorhynchus raphiodon, Applegate, 1970, p. 395, fig. 178A–C.

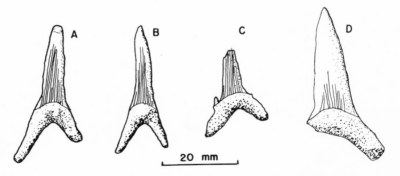

FIG. 12.   *Scapanorhynchus* from Alabama. A, B, C: *S. raphiodon*, anterior teeth. D: *S. rapax*, anterior tooth. After Applegate, 1970. Scale bar 20 mm.

Perhaps the most spectacular of the Alabama Cretaceous sharks, *Scapanorhynchus raphiodon* is also common and easily identified. The teeth are remarkably long and slender with a pronounced sigmoid flexure in side view. The cutting edges are razor-sharp on both sides, and even after ninety million years, the points are often sharp enough to do damage to the careless handler. There is one pair of accessory cusps, which are small, slender, and sharp. These are missing in most specimens.

The labial face of the main cusp bears distinct striae that reach nearly to the tip of the cusp. This feature will readily distinguish

S. *raphiodon* from the very similar S. *rapax* in which the striae extended less than halfway up the main cusp. Teeth of S. *raphiodon* commonly are 2 inches long.

Both species of *Scapanorhynchus* are very similar to, and may be confused with, *Odontaspis macrota striata* (Winkler). The latter species has a very similar morphology, especially similar to S. *raphiodon*. However, O. *macrota striata* shows much finer striations on the lingual face of the crown than does either species of *Scapanorhynchus* (two to three per millimeter rather than one per millimeter). *Scapanorhynchus* is confined to Cretaceous rocks, while O. *striata* is found in the Paleocene and Eocene.

Scapanorhynchus rapax (Quaas) (figs. 12D, 13)

Scapanorhynchus rapax, Applegate, 1970, p. 396, fig. 178D.

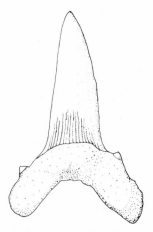

FIG. 13.   *Scapanorhynchus rapax* (Quaas). Lower left anterior tooth. Tombigbee Sand, Perry County, Alabama. Collection of Michael Szabo. x2.

This species is very similar to *Scapanorhynchus raphiodon*. The teeth tend to be somewhat larger, and the main cusp is stouter. The striae on the lingual face extend less than halfway up the crown, thus readily distinguishing it from S. *raphiodon*. It is not so common as S. *raphiodon* but occurs in the Eutaw and the Mooreville, according to Applegate (1970). The species is otherwise known only from the Upper Cretaceous of North Africa.

There is also a tendency for the crown to be even more slender than that of S. *raphiodon*, but this is not a constant character. Otherwise, *Odontaspis macrota* shows the same slender sigmoidal main cusp and a generally similar morphology. It is, however, a much later form, confined to the Tertiary. Small specimens come from the Paleocene of Europe. The species is known in Alabama from the Gosport Sand (White, 1956), and it seems to be the most abundant large shark in that unit. It is not known to be present in the Jackson (upper Eocene), although it appears in rocks of this age elsewhere.

Tuomey (1858A) reports *Lamna elegans* from the Cretaceous of Alabama. This species (from Agassiz) usually has been considered a synonym of *Odontaspis macrota*, which does not occur in the Cretaceous. His report probably is based on specimens of *Scapanorhynchus ráphiodon*, which is easily confused with *O. macrota*. Tuomey (1858A) also reports *Lamna raphiodon* from the Tertiary of Alabama, probably also based on *O. macrota*, as *Lamna raphiodon* is confined to the Cretaceous (*=Scapanorhynchus raphiodon*).

This species has a very wide distribution in Upper Cretaceous rocks, having been reported from Europe, Africa, Indonesia, Japan, Australia, New Zealand, and the Caribbean. In the United States it is known from Alabama, Texas, Kansas, and New Jersey (Applegate, 1970). It also has been the source of numerous erroneous records of other species from Alabama. Woodward's (1889D, p. 365) record of *O. elegans* from the Alabama Eocene is based on mislabeled material of this species from the Alabama Cretaceous. Tuomey's (1858A, p. 163) report of *L. elegans* (the same species) is probably also *S. raphiodon*, as several of his other records may be. The exact identity of Tuomey's material never will be known because records were destroyed when the University of Alabama was burned in 1865.

Odontaspis macrota (Agassiz) (fig. 14)

Lamna (Odontaspis) macrota Agassiz, 1843B, vol. 3, p. 373, pl. 32, figs. 29–31.
Odontaspis macrota Leriche, 1942B, p. 44; White, 1956, pp. 147–148.

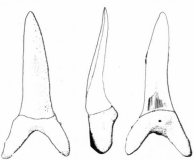

FIG. 14.   *Odontaspis macrota striata*, var. *semistriata* (Leriche). BSC 770, near Monroeville, Monroe County, Alabama. Buccal (outer), anterior (front), and lingual (inner) views. Crown only shown in outline with striae in second and third views. Forms with longer striae (more than ½ crown length) are *O. macrota striata* in a strict sense. x1.

Although strikingly similar to *Scapanorhynchus raphiodon*, *Odontaspis macrota* has four characters that will help to distinguish it:

1. The roots are less massive than those of *S. raphiodon* and smaller in relation to the size of the crown.

2. The median groove in the root stops at the farthest projection of the lingual shelf, rather than continuing almost to the base of the crown as in *S. raphiodon*.

3. The accessory cusps are minute, poorly formed, and often not preserved.

4. The striations are finer than in *S. raphiodon*.

Odontaspis macrota semistriata (Leriche) (fig. 14)

Odontaspis macrota L. Agassiz, praemut. *striata* Winkler, var. *semistriata* Leriche, 1942B, pp. 13–14, pl. 1, figs. 6–8.

This form is only doubtfully distinct from European and American material referred to *Odontaspis macrota striata*. However, as described by Leriche (1942B), there are consistent differences that distinguish this shark from both typical *O. macrota* and *O. macrota striata*. This is perhaps best expressed by elevation to subspecific rank.

The crown is extremely slender and strongly sigmoid in lateral view, although not so strongly as in *Scapanorhynchus*. The accessory cusps are set far down on the sides of the root, which is less widely spread than in *O. macrota striata* (and in *O. macrota*). The striation of the lingual face is much less pronounced and does not extend more than halfway up the main cusp.

In general, these differences are at least as pronounced as those between *Scapanorhynchus raphiodon* and *S. rapax*. If more material indicates complete consistency of these variations, full specific rank seems appropriate.

This shark, one of two species reported thus far from the Paleocene of Alabama, was found at Shoal Creek, Wilcox County. The original specimens are from Texas.

Odontaspis hopei (Agassiz) (fig. 15)

Lamna (Odontaspis) Hopei Agassiz, 1843B, vol. 3, p. 293, pl. 37*a*, figs. 27, 28, 30 (not fig. 20, *fide* Leriche, 1942B, p. 28).

?*Odontaspis cuspidata*, Woodward, 1889D, p. 371 (not of Agassiz; in part this species, *fide* White, 1956, p. 125).

Odontaspis cuspidata praemut. *Hopei*, Leriche, 1942B, p. 28.

Odontaspis (Synodontaspis) hopei, Casier, 1946, pp. 64–66, pl. 2, fig. 11a–b.

Odontaspis hopei, White, 1956, p. 125.

FIG. 15.   *Odontaspis hopei* (Agassiz). After Casier, 1947, from the Ypresian of Belgium. x1.

White (1956) reported this species from Alabama but did not figure his material. Generally, the species is similar to *Odontaspis macrota* in form but smaller and without striae on the lingual face.

Thus far, *O. hopei* is reported only from the Jackson Group of Clarke County, but it also occurs in the Gosport Sand of Choctaw County, west of Silas.

Odontaspis malletiana White (fig. 16)

Odontaspis malletiana White, 1956, pp. 130–131, figs. 20–21, pl. 11, figs. 2–3.

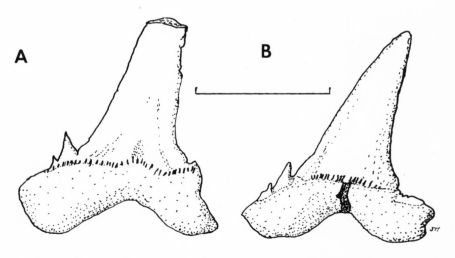

FIG. 16.   *Odontaspis malletiana* White. A: broken right upper anterior lateral, buccal view. B: left upper lateral, lingual view. Holotype. After White, 1956, figs. 20–21. Scale bar 10 mm.

This species is known only from the Jackson Eocene of Clarke County, Alabama (if White's ideas as to the locality of his specimens prove correct). It seems quite distinct from other species of the genus. The crown is in the form of an inclined triangle with two pairs of slender, sharp accessory cusps. The base of the crown shows fine pleats of the enamel on both sides of the tooth. As White (1956) notes, this feature is normal on the buccal face of *Odontaspis*, but on the lingual face it is unique to this species.

At present *O. malletiana* is known only from two teeth, both in the British Museum (Natural History). Neither tooth is complete, though both show the characters of the species. Obviously, there is much to learn about this species, named for the University of Alabama chemist who did the first work on atomic weights.

Odontaspis sp. cf. *O. rutoti* (Winkler)

Odontaspis cf. *rutoti* (Winkler), White, 1956, p. 148.

White (1956) refers only one tooth to this species, noting that it resembles a lower posterior tooth of Winkler's species from Europe but is not typical. White wrote, "This is a small upright tooth, 1 cm. high and the same across the roots. The enamel is smooth on both faces, except for fine puckering at the base of the outer face, where the base is straight and tends to overhang the root. The inner face of the root is strongly protuberant."

Odontaspis ?verticalis (Agassiz) (fig. 17)

Lamna (Odontaspis) verticalis Agassiz, 1843B, vol. 3, p. 294, pl. 37a, fig. 31.
Odontaspis verticalis Casier, 1946, p. 70, pl. 2, fig. 9a–d.
Odontaspis ?verticalis Ag., White, 1956, pp. 131–132.

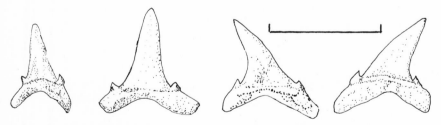

FIG. 17.   *Odontaspis ?verticalis* (Agassiz). After Casier, 1946. fig. 9a–d (Ypresian of Belgium). Scale bar 10 mm.

Collectors should be on the lookout for this species. White (1956) based his identification on a single broken tooth and doubted that it came from Alabama. More material is needed to clarify the relationships of this form, if it does occur in the Jackson Group of Alabama. Our figures are based on material from Belgium. The blade is broad, resembling some species of *Lamna,* and the accessories are weak.

FAMILY LAMNIDAE
mako, white, and mackerel sharks

Shark families often are difficult to distinguish on the basis of teeth. For all practical purposes, the procedure is to decide to what genus and species it should be referred, then decide what modern shark is closest to it and carry the taxonomy upward from there.

The Lamnidae (Isuridae or Oxyrhinidae of some authors) is one of these families. In general, the teeth are simple and triangular, without powerful lateral cusps and serrations. Neither generalization persists on closer examination.

The family is built around three genera. *Lamna,* including the living mackerel sharks, has teeth much like those of *Odontaspis* with a prominent lingual shelf. The median groove in this shelf is not so strongly developed in *Lamna* as in *Odontaspis.* Accessory cusps always are present and are much more prominent than those of most species of *Odontaspis;* they are commonly blunted. However, there is a tendency in paleontological work to confuse these two genera; most species now placed in one genus have been in the other at some time in their nomenclatural histories.

Isurus, typified by the living mako sharks, is much like *Lamna* except for the total absence of accessory cusps. *Isurus* is termed *Oxyrhina* by many workers, particularly in Europe. We will follow the usage of most North American students.

Carcharodon includes the living white shark, *C. carcharias,* the feared "man-eater" of legend and of fact. Up to 18 feet in length, this ravenous hunter is probably responsible for more human deaths than are all other shark species combined. It is extremely pugnacious, even making unprovoked attacks on small boats. Any prey weighing less than 100 pounds is swallowed almost whole, but *Carcharodon* is quite capable of dismembering larger prey.

The teeth of *Carcharodon* are a sight almost as chilling as the entire animal. Great triangular blades with powerfully serrated cutting edges, the teeth attain a length of 3 inches or more in living species. A Miocene species, *C. megalodon,* has teeth reported to reach 8 inches in length, thus representing at least 40 feet of concentrated terror. It is quite probably the most dangerous animal ever to have lived in our seas. Its smaller living relative almost certainly holds that dubious distinction today.

Fortunately for swimmers, the lamnids are generally open-sea sharks. *Lamna* is found commonly in nearshore waters, but it is not particularly dangerous. The other two genera favor the high seas, making only occasional forays into coastal waters. Their geologic range is Cretaceous to Recent. All three genera have fossil representatives in Alabama rocks.

Lamna appendiculata (Agassiz) (fig. 18)

Otodus appendiculata Agassiz, 1843B, vol. 3, p. 270, pl. 32, figs. 1–25.
Lamna appendiculata Applegate, 1970, pp. 396–397, fig. 179E–H.

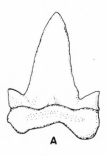
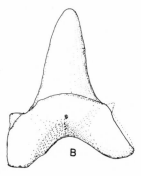

FIG. 18. *Lamna appendiculata* (Agassiz). A: anterior upper tooth, buccal (outer) view. B: anterior lower tooth, lingual (inner) view. Both BSC 769, Sawyerville, Hale County, Alabama. x2.

Common in the Eutaw and Selma rocks of Alabama, this species occurs in Upper Cretaceous strata around the world. Applegate (1970) notes it from Asia, Australia, New Zealand, north and west Africa, Canada, Kansas, New Jersey, and Alabama. The accessory cusps are prominent and triangular (with the tips worn down in most specimens). The root is broad and not deeply bifurcated (forked). The main cusp is quite smooth and erect or slightly inclined and is tallest and most acute in anterior teeth. In the most posterior teeth the main cusp is scarcely larger than the accessories.

Lamna mediavia Leriche (fig. 19)

Lamna mediavia Leriche, 1940A, p. 590 (nomen nudum).
Lamna mediavia Leriche, 1942B, pp. 14–15, 19, pl. 1, figs. 12–19.

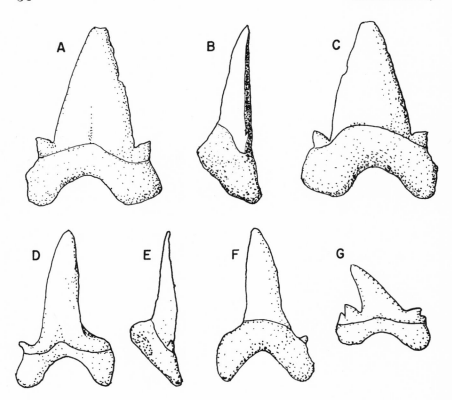

FIG. 19.   *Lamna mediavia* Leriche. A, B, C: large anterior upper right tooth, external,
           lateral, and internal views. D, E, F: anterior lower left tooth, same views. G:
           lateral right upper tooth, external. After Leriche, 1942. All x1.

Leriche (1942B) reports this species from the Midway of Wilcox
County, and from the Lance Formation of Wyoming. The tooth is
highly distinctive among the American sharks, especially in its lateral
cusps. There is one pair, large, triangular, and distinctly inclined away
from the main cusps. A few specimens show two pairs. The main cusp
is erect to strongly inclined, even curved.

The name is a pun on the name of the Midway Group.

Isurus mantelli (Agassiz) (fig. 20)

Oxyrhina mantelli Agassiz, 1843B, vol. 3, p. 280, pl. 33, figs. 1–5.
Isurus mantelli Applegate, 1970, pp. 397–398, fig. 178I–KX.

This species also occurs in Alabama in the Eutaw Formation and the
Selma Group of Late Cretaceous age. As in all species of *Isurus*, there
is no trace of lateral cusps. The crown is tall and heavy. In anterior

FIG. 20.   *Isurus mantelli* (Agassiz). Upper tooth, buccal (outer) view. BSC 771, near
Sawyerville, Hale County, Alabama. x1.

teeth, the slope of the edges of the crown is continued in the root,
giving an almost completely triangular tooth. In more lateral teeth, the
main cusp is considerably narrower than the root, and a pair of accessory carinae (keels or ridges) runs outward almost to the ends of the
root.

Isurus praecursor americanus (Leriche) (fig. 21)

Oxyrhina praecursor var. *americana* Leriche, 1942B, pp. 45–46, pl. 3, figs. 6–13.
Isurus praecursor americana White, 1956, p. 125.

FIG. 21.   *Isurus praecursor americanus* (Leriche). Anterior upper tooth, buccal (outer)
view. BSC 773, Jackson Gosport, near Melvin, Choctaw County, Alabama. x1.

Like *Isurus mantelli*, this is a large species with adult teeth well
over an inch in length. In general, *I. praecursor americanus* has a
more slender crown than that of *I. mantelli*, and the roots of upper
teeth are broader and heavier with much less of a median indentation
in the root. Indeed, the base of the root may be almost straight. The
crowns of both uppers and lowers of *I. praecursor americanus* tend to
be thicker proportionately than those of *I. mantelli*.

Leriche (1942B) reports this subspecies from Alabama, Mississippi,
and South Carolina. All his figured specimens are from the Old Cocoa
Post Office, Choctaw County, Alabama, which is to be considered the
type locality. No type specimen has been designated in print, but the
types are in the collections of the United States National Museum.

56   Chondrichthyes

The change in name is made so that the gender of the subspecies will
agree with that of the genus to which White (1956) transferred it.

In Alabama this species seems to be confined to the upper Eocene
Jackson Group. No specimens appear to have been collected from the
Claiborne Group, despite considerable material from these rocks.

Carcharodon (Procarcharodon) angustidens Agassiz (fig. 22)

Carcharodon angustidens Agassiz, 1843B.
Carcharodon auriculatus Woodward, 1889D, p. 413.
Carcharodon angustidens L. Agassiz, praemut. cf. *Sokolowi* Jaekel, Leriche, 1942B,
    pp. 46–47, pl. 3, figs. 1–5.
Carcharodon angustidens White, 1956, p. 125.

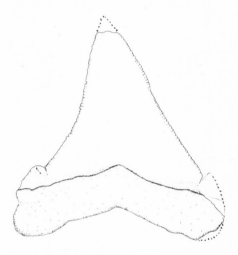

FIG. 22.   *Carcharodon (Procarcharodon) angustidens* Agassiz. BSC 122, locality un-
           known. Buccal (outer) view. x1.

Despite some difficulties in naming, this shark is the most easily
recognized Alabama lamnid, and the largest. Two-inch teeth are runts;
twice that size, and you are beginning to find remarkable ones. The
teeth are tall triangular blades, erect or slightly inclined. Both edges
are powerfully and coarsely serrated. There are distinct lateral cusps,
also serrated.

Early reports of the even larger *Carcharodon megalodon* from Alabama are based either on this species or on properly identified specimens that were mislabeled. For example, Woodward's (1889D, p. 418)
specimens of *C. megalodon,* though labeled "Alabama," were shown
by White (1956, p. 126) to be from Malta.

The teeth of *C. angustidens* differ from those of *C. megalodon* in several respects; *C. megalodon* has much finer serrations, no accessory cusps, and a broader crown.

FAMILY ANACORACIDAE

This extinct family includes a primitive group of true sharks that, in many ways, paralleled the tiger sharks *(Galeocerdo)* in tooth development. The crowns are always strongly inclined triangles, usually serrated. Posterior to the main cusp is a low cutting ridge, also usually serrated. Anacoracids are known from Cretaceous through Miocene but are most abundant in the Upper Cretaceous. In Alabama, they are known only from rocks of the latter age.

Squalicorax falcatus (Agassiz) (fig. 23)

Corax falcatus Agassiz 1843B, vol. 3, p. 226, pl. 26, fig. 14, pl. 26a, figs. 1–15 (latter as *Galeus pristiodontus*).
Corax falcatus, Tuomey, 1858A, p. 261.
Squalicorax falcatus, Applegate, 1970, pp. 393–395, figs. 176A–F, 178L, N, 179B.

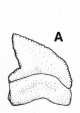 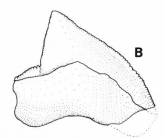 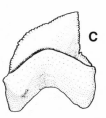

FIG. 23.   *Squalicorax falcatus* (Agassiz). All lateral teeth. A, B: BSC 772, near Sawyerville, Hale County, Alabama. A: left upper, buccal (outer) view. B: right upper, lingual (inner) view. C: BSC 698, locality unknown, right lower, lingual view. x2.

This species has long been known from Alabama. Applegate (1970) describes not only teeth but also vertebrae and dermal denticles of *Squalicorax falcatus* from the Mooreville Chalk of Alabama. As the identification of both vertebrae and dermal denticles is a highly specialized subject, and as the latter are unlikely to be recovered by any but the most exacting collecting, only teeth will be discussed here.

Teeth of *S. falcatus* have a strongly convex and serrated anterior edge. Inclination of the main cusp increases posteriorly while the convexity of the anterior edge decreases toward the rear of the mouth. In general, teeth of *Squalicorax* can be distinguished readily from those of *Galeocerdo* by two features. The roots of *Galeocerdo* teeth are very shallow, and the posterior cutting edge is strongly serrate, even denticulate. *Squalicorax* roots are deep and broad, often as deep as half the height of the crown or more, and the posterior cutting edge is no more strongly serrated than the edge of the main cusp. Often the serrations here are weaker than those on the main cusp.

The species is confined to the Upper Cretaceous but has a nearly worldwide distribution.

Squalicorax pristiodontus (Agassiz) (fig. 24)

Galeus pristiodontus Agassiz, 1843B, vol. 3, p. 224, pl. 26, figs. 9–13.
Galeus pristiodontus, Tuomey, 1858A, p. 261.
Squalicorax pristiodontus, Applegate, 1970, p. 396.

FIG. 24.   *Squalicorax pristiodontus* (Agassiz). Tooth, 4 miles south of Sawyerville, Hale County, Alabama. Basal Mooreville Chalk. BSC collection. x2.

Reported twice from Alabama, this species was first recorded in Tuomey's (1858A) faunal list, and again by Applegate (1970). Woodward (1889D) reported this Cretaceous species from the Eocene of Alabama on mislabeled specimens from the Enniskillen collection (White, 1956). Several later authors, including Hay (1902A), have persisted in the use of Woodward's erroneous reference.

Applegate (1970) wrote that *Squalicorax pristiodontus* is distinguished from *S. falcatus* by "the large size and the broad low crown of the teeth." The anterior edge is much less arched than in *S. falcatus.* Specimens are known in Alabama from the Tombigbee Sand Member of the Eutaw Formation and the Mooreville and the Demopolis.

Pseudocorax affinis (Agassiz)

Corax affinis Agassiz, 1843B, vol. 3, p. 227, pl. 26, fig. 2.
Pseudocorax affinis, Applegate, 1970, p. 395, figs. 177, 178M.

This shark is a very small species with teeth about 5 millimeters long. As such, it is not likely to be found by surface collecting unless by very intensive work or by accident. The tooth has a strongly inclined main cusp with a straight or slightly concave anterior edge. The posterior carina is separated from the main cusp by a prominent notch. Serrations are weak or absent. The species is known from the Upper Cretaceous of Europe, North Africa, and North America.

The only species in the Alabama fauna that might be confused with *Pseudocorax affinis* is *Rhizoprionodon* sp. from the Gosport, which has a very similar crown to that of *P. affinis*, but the root of *Rhizoprionodon* is much shallower and has a prominent median notch.

FAMILY ALOPIIDAE
thresher sharks

Living thresher sharks are unmistakable. The upper lobe of the tail is enormously elongated, often as long as the rest of the body. These are open-sea forms, living in surface waters. Only one genus, *Alopias*, is included in the family, and it is known from Eocene to Recent. The Alopiidae is sometimes combined with either the Lamnidae or the Odontaspidae.

The teeth of *Alopias* never are serrated, nor do they possess accessory cusps. The root is very shallow, and the crown always seems to appear worn, without the sharp cutting edges seen in most other sharks. Only one form is known in the Alabama fossil record.

Alopias latidens alabamensis White (fig. 25)

Alopias latidens (Leriche) *alabamensis* White, 1956, p. 132, figs. 28–36, pl. 11, figs. 5–6.

The teeth of this subspecies are tall triangles, with a slight to pronounced inclination. The crown trails off posteriorly into a poorly defined carina. The root is very shallow and small in relation to the crown. There is a triangular depressed area at the base of the main cusp on the labial face. Thus far, *Alopias latidens alabamensis* has been reported only from the Jackson Eocene (White, 1956), but it is common in the Gosport Sand (middle Eocene) at Little Stave Creek. The teeth are about a centimeter in both width and height.

FAMILY SCYLIORHINIDAE
cat sharks, some "dogfish"

This is a fairly primitive family of bottom-dwelling sharks. Most species are small, averaging between 1 and 2 feet in length. The tooth roots have a deep, sharp-sided median groove, and the crowns have one or more pairs of accessory cusps.

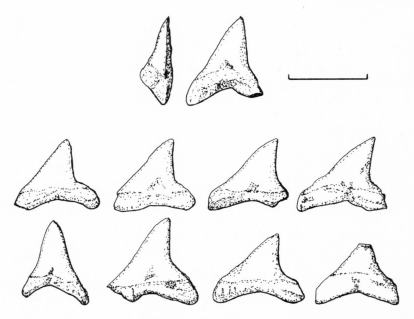

FIG. 25. *Alopias latidens alabamensis* White. Teeth, Choctaw County, Alabama. After White, 1956. Scale bar 1 cm.

Scyliorhinus enniskilleni White (fig. 26)

Scyliorhinus enniskilleni White, 1956, pp. 128–130, figs. 1–19, pl. 11, fig. 1.

The only fossil cat shark from Alabama is a notable exception to the general small size of the family representatives. The teeth reach 1.5 centimeters in height, representing roughly 8 feet of shark. Most specimens, however, are about half this size. This length is still twice the average for the family.

At first glance, the teeth of *Scyliorhinus enniskilleni* resemble those of the more slender-toothed species of *Lamna* and *Odontaspis*—tall, sigmoid main cusps with one pair of acutely triangular lateral cusps. The main cusp is crudely fluted on the lingual face, reminiscent of *O. macrota*. The root, however, is totally different. Rather than being long and deeply forked, the roots of *Scyliorhinus* are short, with a nearly flat, beveled basal surface. This base is split by a sharp-sided groove.

1a 1 2 3 4 5 6 7
8a 8 9 10 11 12 13
14 15 16 17 18 19
Scyliorhinus enniskilleni, n. sp.
Figs. 1-9. Upper teeth. 1,3,6,9, from left side, remainder from right. Holotype, fig. 4. (P.30633, P.30534, P.30614, P.30634, P.30611, P.30505, P.30576, P.30615, P.30577). 10-19. Lower teeth. 10,13,15-17. From right side, remainder from left. (P.30637-9, P.30613, P.30575, P.30616, P.30631, P.30636, P.30635, P.30632). 1,6,13,15-18. Show inner face. 1a,8a. Front view. All teeth from upper Eocene, Clarke County, Alabama. X 1 1/3.

FIG. 26. *Scyliorhinus enniskilleni* White. After White, 1956. Scale bar 3 cm.

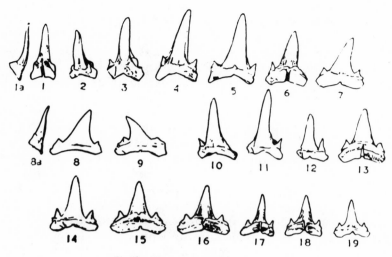

Scyliorhinus enniskilleni was first reported (White, 1956) from the Jackson Eocene of Clarke County, Alabama. Remains of the species also are abundant in the Gosport Formation in Clarke and Monroe counties. The species thus far is known to range from middle to upper Eocene and has been reported only in Alabama.

FAMILY CARCHARHINIDAE
requiem sharks

The largest and most complex living shark family, Carcharhinidae also seems to be the most advanced, at least among the galeoids. The family contains some familiar forms, such as the tiger shark *(Galeocerdo)* and the blue shark *(Prionace).* Most of the genera are poorly known, and many lack any popular names. Most genera are living, but two obscure forms are totally extinct. One other genus,

Hemipristis, is abundant in the fossil record, but only one living specimen has ever been caught, over a century ago in the Red Sea.

The common name of the family, the requiem sharks, conjures up visions of tolling funeral bells. With the exception of *Carcharodon* (Lamnidae), the two most dangerous shark genera in modern waters belong to this family. *Galeocerdo*, the tiger shark, is responsible for many attacks on swimmers near beaches. Well-documented attacks by the blue shark, *Prionace*, are less common, because the genus frequents deeper, offshore waters. Its reputation is bad among sailors who have more chances of meeting it in unfavorable circumstances.

This family is too varied to characterize easily on the basis of teeth. The root is very shallow and resembles that of the Scyliorhinidae. The main cusp usually is inclined, and accessory cusps normally are developed only posterior to it. The teeth often are serrated. For some reason, the most common living genus, *Carcharhinus*, has a very poor fossil record. The family appears in the Upper Jurassic but does not become abundant until the Upper Cretaceous. All known Alabama representatives are from the Eocene.

Galeocerdo clarkensis White (fig. 27)

Galeocerdo clarkensis White, 1956, pp. 145–146, figs. 24–26, pl. 11, figs. 12–14.

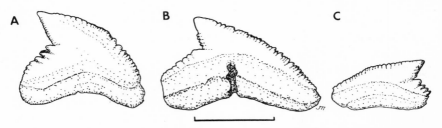

FIG. 27.   *Galeocerdo clarkensis* White. A: anterior tooth, buccal view (holotype). B: lateral tooth, lingual view. C: posterior lateral tooth, buccal view. After White, 1956, figs. 24–26. Scale bar 10 mm.

This tiger shark is fairly typical. The main cusp is low, particularly on posterior teeth. It is strongly inclined, and the anterior edge is powerfully serrated, with the serrations largest near the base of the crown. Serrations on the posterior edge are much smaller. Posterior to the main cusp is a series of accessory cusps that diminishes in size away from the main cusp, sometimes being little more than serrations posteriorly. As in other tiger sharks, the middle part of the root rises much higher on the lingual face of the crown than on the labial face.

This species is closely related to *Galeocerdo alabamensis* Leriche. The leading edge of the main cusp in *G. clarkensis* is smoothly convex, while in *G. alabamensis* it is straight, or even concave or S-shaped. So far *G. clarkensis* is reported only from the Jackson, but it is common in the Gosport and doubtfully present in the Tallahatta (middle Eocene).

Galeocerdo alabamensis Leriche (fig. 28)

Galeocerdo alabamensis Leriche, 1942B, p. 48, pl. 4, fig. 2.

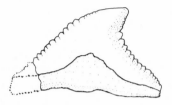 

FIG. 28. *Galeocerdo alabamensis* Leriche. Left: right upper tooth, buccal view, BSC 774, from near Burnt Corn, Monroe County, Alabama. Right: right lower posterior tooth, lingual view, BSC 775, near Peterman, Monroe County, Alabama. Both x2.

Much like *Galeocerdo clarkensis* White, this species differs in the form of the anterior edge of the main cusp (see above). Leriche (1942B) recorded it from the Jackson, but it is also present in the Tallahatta near Peterman, Monroe County, and in the Gosport at Little Stave Creek, Clarke County.

Two other species of *Galeocerdo* have been reported from the Eocene of Alabama. Woodward (1889D, p. 443, 446) noted *G. contortus* Gibbes and *G. aduncus* Agassiz from Clarke County. White (1956) interprets some of these specimens as representing *G. clarkensis* and others, originally correctly identified, as being mislabeled specimens from Malta. Earlier, Tuomey (1858A) recorded *G. aduncus* and *G. latidens* in the Tertiary of Alabama. As Tuomey's specimens have not survived, these specimens are assumed herein to have belonged to one of the species described above.

Hemipristis wyattdurhami White (fig. 29)

Hemipristis wyattdurhami White, 1956, pp. 133–317, figs. 40–47, pl. 11, fig. 4.

The name of this genus can be translated as "half-saw." This designation is appropriate because only the posterior edge of the tooth crown is regularly serrated. There may be a few serrations, very crude and irregular, at the base of the anterior edge. A peculiar feature of the genus is that each tooth has an unusually large pulp cavity, which

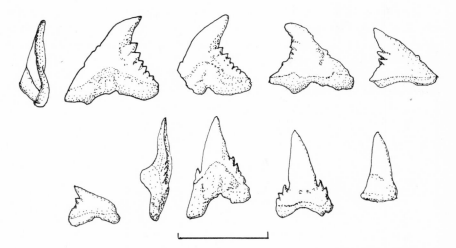

FIG. 29. *Hemipristis wyattdurhami* White. Upper Eocene, Clarke County, Alabama.
Upper row, left to right: upper teeth, two views of holotype, an upper right
lateral and three left laterals. Lower row, left to right: lower teeth, posterior,
two laterals, anterior (crown only). After White, 1956, figs. 40–47. Scale bar
10 mm.

appears on the outside as a stained area extending up for some distance into the crown above the root.

The main cusp of *Hemipristis wyattdurhami* is tall, usually curved
and inclined, and smooth on both margins except for occasional vestigial serrations on the base of the anterior edge. Posterior to the main
cusp, and beginning about a third of the way down from the tip, is a
series of accessory cusps, diminishing posteriorly.

This species is markedly different from *H. serra* Agassiz, a Miocene
species and the only other species of the genus common in eastern
North America. In *H. serra* the main cusp is much broader, and the
posterior denticles begin as serrations almost at the tip and increase in
size toward the base.

The most recognizable teeth of *H. wyattdurhami* are uppers, on
which the above description is based. Anterior lowers have the main
cusp almost erect, with accessory cusps, usually more numerous posteriorly, confined to the base.

Physodon secundus (Winkler)

Trigonodus secundus Winkler, 1876A, p. 5, pl. 1, figs. 4–5.
Physodon secundus, Casier, 1946, p. 91; 1967, p. 29, pl. 7, figs. 18–20; White, 1956, pp.
144, 148.

This is a peculiar and easily recognized form, but so far it has eluded the authors' collecting. The root is flat-based with a strong median groove. The main cusp (there are no accessories) is quite slender and strongly inclined, appearing somewhat bent. Its tip is noticeably upturned. White (1956) reports the species from both the Gosport and Jackson.

Aprionodon greyegertoni (White) (fig. 30)

Hypoprion greyegertoni White, 1956, pp. 137–139, figs. 48–56, pl. 11, fig. 7.

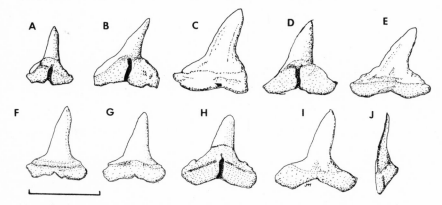

FIG. 30.   *Aprionodon greyegertoni* (White). Upper row, from left side: A, B, C, left uppers from lingual side; D and E, right uppers from buccal side; E is holotype. Lower row: H is a left upper in lingual view; F, G, I, and J are left lowers in buccal view; J is posterior view of I showing sigmoid curvature. After White, 1956, figs. 48–56. Scale bar 10 mm.

The teeth of this species are quite simple in appearance. The main cusp is slender and gently sigmoid, rising from a flaring base on the buccal side of the root. It is noticeably, but not strongly, inclined. The base of the root is "dead flat . . . divided by a well-marked vertical canal" (White, 1956, p. 137). Accessory cusps are lacking; the basal flare of the crown may be wavy or weakly notched, but there are no true serrations.

There is some difficulty in separating this species from *Negaprion gibbesi gilmorei*, the next species described. At first glance, they appear quite similar. However, teeth of *Aprionodon greyegertoni* are taller than broad; the reverse is true of most *N. gibbesi gilmorei*. The latter species does not have the median groove ("vertical canal") so sharply marked as in the former. Worn teeth, particularly if rootless, may be unidentifiable.

The generic assignment of this species is the only taxonomic point on which we will depart from White's (1956) treatment of the Eocene fishes of Alabama. We do so on both morphologic and ecologic grounds. Living species of *Hypoprion* are characterized by having accessory cusps on the upper teeth, usually two on the posterior edge (Bigelow and Schroeder, 1948, p. 315ff.). On the other hand, teeth of *Aprionodon* are very similar to White's figures. Ecologically, *Hypoprion* is a deep-living shark, caught only at night in Cuban waters at depths greater than 100 fathoms (Bigelow and Schroeder, 1948). Poll (1951, p. 46) records the genus off West Africa at depths of 100 to 310 meters. The Alabama rocks in which "*Hypoprion*" *greyegertoni* is found seem to represent shallower waters. *Aprionodon* is a nearshore form (Bigelow and Schroeder, 1948, p. 303ff.; Poll, 1951, p. 154), further suggesting referral of the species to this genus.

Negaprion gibbesi gilmorei (Leriche) (fig. 31)

Galeocerdo minor, Gibbes, 1849A, p. 192, pl. 25, figs. 63–65 (not of Agassiz, 1843B).
Oxyrhina minuta, Gibbes, 1849A, p. 192, pl. 27, fig. 164 (not of Agassiz, 1843B).
Carcharias (Aprionodon) gibbesi Woodward (1889D, p. 438) (in part).
Sphryna gilmorei Leriche, 1942B, p. 47, pl. 4, fig. 1.
Negaprion gibbesi gilmorei, White, 1956, pp. 139–144, figs. 57–76, pl. 11, fig. 10.

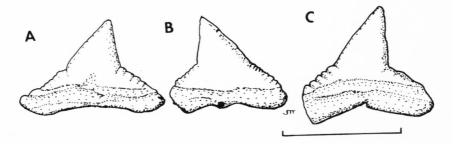

FIG. 31.   *Negaprion gibbesi gilmorei* (Leriche). Three large upper anterior lateral teeth, buccal view. A, C: right, B: left. After White, 1956, figs. 22, 23, 27. Scale bar 10 mm.

White's opening remark on this form is appropriate: "The history of this species is certainly curious" (1956, p. 139). For a review of its taxonomic peregrinations, see White (1956).

The tooth is fairly ordinary looking to have stirred up so much turmoil. The main cusp is triangular, erect to slightly inclined, and not serrated. It flares basally into a pair of broadly spread ridges, which may be smooth, wrinkled, or even faintly denticulate. In addition to the distinctions discussed above under *Aprionodon greyegertoni*, *Negaprion gibbesi gilmorei* is somewhat smaller (maximum 12 millimeters broad and 10 millimeters high as against 16 millimeters). It is known only from the Eocene of Alabama where it is abundant in the Jackson and Gosport.

Galeorhinus recticonus claibornensis White (fig. 32)

Galeorhinus recticonus (Winkler) *claibornensis* White, 1956, p. 148, fig. 97, pl. 11, fig. 11.

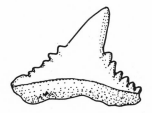

FIG. 32.   *Galeorhinus recticonus claibornensis* White. After White, 1956, pl. 11, fig. 11. x4.

The highly distinctive tooth of this shark is fairly small (5 to 10 millimeters in larger specimens). The crown is a broad triangle, only gently inclined, with a slowly flaring base. On each side of the main cusp are relatively large denticles, the most distinctive character of the species. These are more numerous than in typical European specimens of *Galeorhinus recticonus*, for which reason White differentiated this subspecies. His single specimen had seven anterior and six posterior denticles. A group of about twenty specimens from the Gosport Formation at Little Stave Creek, Clarke County, shows some variation in this count. There are five to eight anterior denticles and five to seven posterior in well-preserved teeth. White's original specimen was from the Gosport at Claiborne, Monroe County.

A more typical specimen of *G. recticonus*, probably referable to *G. recticonus recticonus*, occurs in the Tallahatta Formation, ¼ mile southeast of Peterman, Monroe County. The lateral denticles are much larger relative to the crown size, and the number on both sides is three to four. These differences from *G. recticonus claibornensis* are exactly those used by White (1956) to distinguish his subspecies from the typical form.

Galeorhinus sp. cf. *G. falconeri* (White)

Galeorhinus cf. *falconeri* (White), White, 1956, pp. 144–145, 148.

White (1956) noted teeth of this species from both the Gosport of Monroe County and the Jackson of Clarke County but figured neither.

The specimens we have seen fit his description well except for size (1.1 to 1.6 centimeters for his, a maximum of 5 millimeters for ours). Our specimens are from the Tallahatta Formation ¼ mile southeast of Peterman, Monroe County, a locality that yields much tiny material but not much in larger sizes. The main cusp is a strongly inclined triangle with an upturned tip. There are two or more posterior denticles, trailing off into a crenulated carina. There may be vague, incipient crenulations near the base of the anterior margin.

Except for the presence of posterior denticles, these teeth strongly resemble those referred below to *Rhizoprionodon*. Typical teeth of *Galeorhinus falconeri* otherwise are known only from the Eocene of Nigeria (White, 1926, 1956).

Rhizoprionodon sp. (fig. 33)

FIG. 33.   *Rhizoprionodon* sp., BSC 776, near Jackson, Clarke County, Alabama. Left is lingual view, others buccal. x2. Similar, usually slightly larger, teeth with distinct crenulations on the posterior blade are *Physodon*.

As far as we can determine, this is the first noted occurrence of this genus in the North American fossil record. It is common in the Gosport Formation at Little Stave Creek, Clarke County, but its small size makes its recovery unlikely by any means but bulk screening.

The main cusp resembles that of *Galeorhinus falconeri*, a broad-based triangle sharply inclined posteriorly. There is a posterior carina, distinctly set off from the main cusp by a sharp notch. This carina may be faintly crenulated but is never denticulate. The tip of the main cusp is distinctly upturned. The largest in our collections has a base 7 millimeters long, and specimens range down to a minute 1.5 millimeter. There seems to be no character to distinguish these teeth from those of the living *Rhizoprionodon terraenovae*.

The genus normally is called *Scoliodon*, but we follow Springer (1964) in the use of *Rhizoprionodon*. The living representative commonly is termed the sharp-nosed shark. This is a small, inoffensive, nearshore shark, frequently caught in bays and around piers and pilings. Bigelow and Schroeder (1948, pp. 295–303) indicate that it is taken in brackish and fresh waters and is common in the surf. They know of no specimens taken in more than a few fathoms of water or more than a mile or two offshore. However, Poll (1951, p. 39) finds it in up to 75 meters of water off West Africa. His data also suggest that the species is confined to greenish or black muddy bottoms, ranging in grain size from muds to muddy sands. This environment fits well with that of the basal Gosport at Little Stave Creek, which is a green clayey sand.

OTHER ALABAMA FOSSIL SHARKS

In the collections of the Geological Survey of Alabama, Birmingham-Southern College, and in recently made collections (as yet unsettled) there are numerous specimens that have thus far defied our best efforts at identification. Probably several new records and perhaps new species are represented in this material. Further collection and study is presently (1980) going on.

Although Applegate (1970) was unable to identify the majority of the shark dermal denticles that he describes and figures from the Mooreville Chalk, they seem to represent a fauna more varied than was indicated by teeth and vertebrae. At least one, perhaps two, smooth dogfishes were represented *(Mustelus* and/or *Triakis)*. Even more startling was the presence of dermal denticles that closely resembled those of the gigantic living whale shark *(Rhincodon)*. Because its teeth are minute, scarcely larger than dermal denticles, the whale shark is all but unknown in the fossil record. These tiny denticles may well indicate an ancient lineage for this titanic form. In general, dermal denticles need further study.

In addition to the possibilities noted above, many sedimentary rock units in Alabama have hardly been touched for fossil sharks, at least for serious work. Some of these are known to have abundant teeth, so there is room for much work.

Order Batoidea
sawfishes, skates, rays

These are flattened, usually bottom-dwelling, cartilaginous fishes, probably descended from the hybodonts. The flattened shape of the body is distinctive in most cases, as are the enormous pectoral fins with which batoids "fly" through water in a stately, measured fashion. A few modified true sharks have similar adaptations; these may be distinguished by the position of the gill slits. The gill slits invariably are on the sides of the body of a true shark (except in *Cetorhinus*, the basking shark, in which they nearly circle the body). In batoids, the gill slits are always on the ventral (under) side of the body.

Most of the fossils of batoids fall into three categories: teeth, dermal denticles, and fin spines. In a few cases, calcified vertebrae are found. Even more rare are complete fossil batoids; these are known from very few places on earth, and none have come yet from Alabama.

The true teeth of batoids are all crushing organs. Most are flat-topped, with a more or less complex root. A few are minute and pointed; together, these form a broad rasping pavement. Dermal denticles appear as common fossils in two groups. The highly ornamented "thorns" in the skin of a skate are greatly enlarged dermal denticles. The rostral teeth (on the "saw") of sawfishes ultimately develop from the same source. Fin spines are most common among the rays. These are the rays' "stingers," long daggers with a row of barbs on both edges, and are difficult, if not impossible, to identify even to family.

Batoids are known as fossils as early in the geologic time scale as the Late Jurassic and are abundant and varied in modern seas.

Suborder Pristoidea
sawfishes

These are among the strangest fishes of modern seas and, therefore, completely unmistakable. The front of the skull is drawn out into a long, flattened blade, the rostrum. On either side of this blade is a row of rostral "teeth" (really enlarged dermal denticles). Their feeding habits are well known. A sawfish will charge into a school of small fish, lay about violently with his rostrum, and then pick up the dead and wounded at his leisure. In this mode of feeding, the rostral teeth are of considerable use to the animal.

There is but one family of sawfishes, the Pristidae, known from Lower Cretaceous to Recent.

FAMILY PRISTIDAE

The living sawfishes have been described above. The oldest speci-mens of the living subfamily, the Pristinae, are of Eocene age. The rostral teeth of this subfamily bear no enamel and are set in deep sockets in the rostrum.

The ancestors of the living sawfishes are to be sought in the subfam-ily Ganopristinae (Sclerorhynchinae of some authors). These range from Early Cretaceous (Thurmond, 1972) to possibly earliest Paleocene (Slaughter and Steiner, 1968) but flourished mainly in the Late Cretaceous. Their rostral teeth had saddle-shaped bases that straddled the sides of the rostrum, and are enameled, at least on part of the crown. Many genera show barbs on the rostral teeth.

It is difficult to reconcile this last feature with the feeding habits of living sawfishes. A slashing attack with the rostrum would not have produced dead and wounded in the water. Rather, the small fishes would have been held on the rostrum by the barbs, out of reach of the mouth. Possibly, this outcome was desired. The prey could then be shaken off one by one and engulfed at leisure. This procedure would be less wasteful than that of the Pristinae, and there would be less chance of scavengers' stealing the prey before it was eaten.

No ganopristines have been reported from Alabama, but an occurrence is cited below. Leriche (1942B) noted a sawfish vertebra from the upper Eocene from southwestern Mississippi, and White (1956) recorded a fragment of rostral tooth from the same unit in Clarke County, Alabama.

Ischyrhiza mira Leidy (fig. 34)

Ischyrhiza mira Leidy, 1856K, p. 221.
Ischyrhiza mira, Slaughter and Steiner, 1968, pp. 234–237, text-fig. 3H-I.

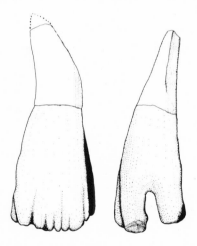

FIG. 34.   Ischyrhiza mira Leidy. Left rostral tooth, dorsal and anterior views. Near Epes, Sumter County, Alabama. D. E. Jones Collection. x1.

Teeth of this species are recognized readily. The crown is long and enameled, without barbs or striae. The root is large, nearly as long as the crown, with a deep groove in the base. The base of the root commonly is fluted. These teeth are mistaken frequently in amateur collections for the teeth of ichthyosaurs.

Pristis sp.

White (1956) extended into the Jackson of Alabama Leriche's (1942B) record of the living genus *Pristis* from rocks of this group in Mississippi. We also have collected fragments of rostral teeth from the Gosport at Little Stave Creek and in Monroe County.

Complete rostral teeth of *Pristis* are quite distinctive. They are long, flattened shafts of dentine with no enamel. Crude flutes are present near the base and may extend to the tip, which is placed near the trailing edge. This trailing edge commonly shows a deep posterior groove (sulcus). In rare cases, fossil rostra may be found, as the rostrum is often calcified.

Suborder Rajoidea

skates

The skates are flattened bottom dwellers often confused with the rays. The tail of a skate is obviously fleshy and part of the body and has two fins. It never is reduced to a slender, whiplike appendage as in the rays.

Skate teeth are exceedingly minute, usually well under $\frac{1}{10}$ inch. It is unlikely that they will be found by surface picking, and only the larger forms are found by screening with a normal window screen. The tooth has a deep root sharply divided by a median groove. The crown is strongly domed and often ornamented with cuspules and ridges.

More common as fossils are the enlarged dermal denticles that are scattered over the dorsal surface of a living skate. These can be nearly an inch across and in shape resemble the conical chocolate chips used in cookies and desserts. The flaring base of the denticle often is ornamented with radiating ridges, and the margin may be scalloped. They are quite sharp in living skates (fossils are usually worn) and are called "thorns."

The oldest skates are from Upper Cretaceous rocks. Most are referred to the genus *Raja*.

FAMILY RAJIDAE

Raja sp.

This record is the first of fossil skates from Alabama. Our specimens are isolated dermal denticles from the basal Gosport at Little Stave Creek. Of the two types found, one has a strongly flaring base with prominent radiating ridges; the other is more conical, without ornamentation. They probably represent two species, one or both of which may be undescribed. The smooth, conical forms are similar to those of the living clearnose skate, *Raja eglanteria,* a nearshore species. The more ornamented specimens are of a type common among the skates.

Suborder Myliobatoidea
rays

The rays are characterized by slender, whiplike tails, extremely flattened bodies, and enormous pectoral fins. The dorsal fins are usually vestigial, but in many genera the dorsal fin spine survives, adapted into a long, dagger-like "stinger" with both edges bearing a row of barbs. Its presence makes a ray underfoot a danger to waders. It inflicts deep wounds and usually breaks off in the flesh. The barbs prevent its being withdrawn without a physician's help, and the slimy covering carries a large bacterial population, often leading to gangrene without prompt medical attention. Such stingers are common as fossils and may be large in size. It is not possible at present to identify the type of ray, even to family, from the stinger. However, very large specimens probably are myliobatids.

Ray teeth are all flattened crushing plates, assembled into a broad grinding surface in the mouth. The principal food of rays is shellfish. The earliest known rays are Early Cretaceous in age.

FAMILY DASYATIDAE
stingrays, "stingarees"

Dating from the Early Cretaceous, this primitive group of rays has numerous flourishing descendants today.

The family is distinguished principally by the teeth, which are small, rhombic or hexagonal crushers. The root usually is deep and always is divided by a single median groove. The crown is nearly flat and never highly domed as in the skates.

Hypolophus sp. (fig. 35)

Even though this genus has not been reported from Alabama previously, its presence in the Upper Cretaceous is no surprise. Our specimens come from the base of the Mooreville Chalk near Sawyerville, Hale County. The tooth is distinguished by its fairly large, pronouncedly hexagonal crown and the nearly vertical sides of the crown.

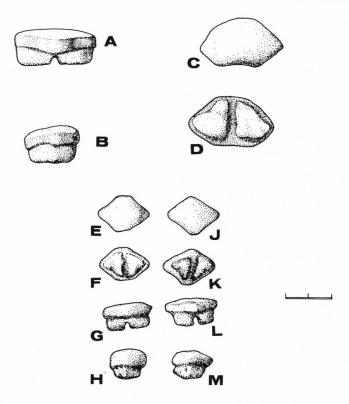

FIG. 35.   *Hypolophus mcnultyi* Thurmond. A–D: holotype, SMUSMP 62208, ?an-
terior, lateral, occlusal, and basal views of a tooth of distorted hexagonal
type. E–M: two teeth from 62209. E–H: tooth of symmetrical hexagonal type,
occlusal, basal, ?anterior, and lateral views. J–M, tooth of rhombic type,
occlusal, basal, ?anterior, and lateral views. From Thurmond, 1971. Middle
Paluxy of Texas. Scale in millimeters.

The root is not so sharply biparted as in other dasyatids and is much
shallower than the crown.

Dasyatis sp.

This genus, as well, has not been reported from Alabama before.
This occurrence is based on teeth and questionable stingers from the
basal Gosport (middle Eocene) at Little Stave Creek. The teeth have a
rhombic crown and a deep, offset, biparted root. The stingers are
referred here on the basis of size. Some or all may represent immature
myliobatids.

FAMILY MYLIOBATIDAE
eagle rays

These large rays are among the most graceful of sea animals. Their pectoral fins are extremely long and pointed and are used for stately "flying" through the water. The head extends in front of the pectorals, rather than being enclosed by these fins as in the dasyatids.

The teeth are quite distinctive, having a multiply divided root instead of the single median groove of the dasyatids. The root often has a comblike appearance. The crown is nearly flat or slightly domed. Median teeth are very broad (from side to side) and short (from front to back) but not so strongly widened as those of the Rhinopteridae. Lateral teeth, if present, are blocky, about as wide as long, and rhombic or hexagonal. They are never strongly widened.

The earliest myliobatids are from the Upper Cretaceous. They are extremely abundant in the Eocene of Alabama.

Myliobatis sp. (fig. 36)

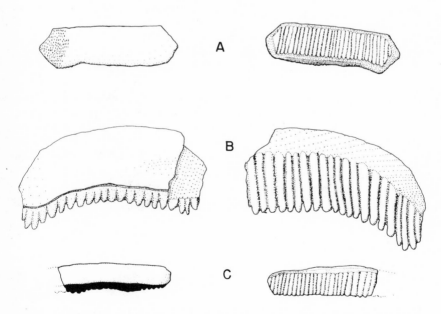

FIG. 36.   Myliobatoid median teeth: left, occlusal view; right, basal view. A: *Myliobatis* sp. BSC 777, from near Burnt Corn, Monroe County, Alabama. B: *Aetiobatis* sp. cf. *A. irregularis* Agassiz. BSC 778, from near Georgiana, Butler County, Alabama. C: *Rhinoptera* sp. BSC 779, from Little Stave Creek, Clarke County, Alabama. B, C: broken, each about 50 percent complete. All x2.

Isolated teeth of this species are the most common vertebrate fossils in the Gosport. Thousands have been found on a single outcrop. They are also abundant in the Jackson and Tallahatta.

The wide variety of shapes present probably indicates several species in the Alabama Eocene. In general, any tooth whose width is less than six times its length is probably of this genus. In contrast to the abundance of isolated teeth, more or less complete toothplates are rare. They will be very important finds because they will permit more precise identification of the species present.

Aetiobatis sp. (fig. 36)
spotted eagle rays

The teeth of this genus are very different from those of *Myliobatis*. They have the same comblike root, but this is greatly slanted posteriorly and shows extensively at the posterior margin of the crown. The root also is bent strongly along the midline, even to the extent of having a chevronlike appearance. There are no lateral teeth. This genus is much rarer than *Myliobatis* in the Alabama Eocene but is found at the same localities.

FAMILY RHINOPTERIDAE
cow-nosed rays

This family is only doubtfully distinct from the Myliobatidae, and many workers combine them. The teeth are basically similar but much more widened, often ten times as wide as long. The laterals, at least the first series, are also strongly widened, and may be wider than the medians. The presence of widened teeth that are not bilaterally symmetrical (and thus not medians) is a dead giveaway for the rhinopterids.

The family is known from the Upper Cretaceous.

Rhinoptera sp. (fig. 36)

Teeth of this genus are scarcely less common than those of *Myliobatis* in the Gosport at Little Stave Creek. See the family for a description.

Subclass Holocephali

Order Chimaeriformes

Suborder Chimaeroidei
chimaeras

The chimaeras, also known as ratfish, are very peculiar cartilaginous fishes, so different from the sharks as to merit separation at the subclass level. Eaters of shellfish, they have a crushing dentition. The jaws are often strongly calcified and are not uncommon as fossils. Related forms are known as far back as the Permian.

FAMILY EDAPHODONTIDAE

This family has massive crushing jaws, often up to 3 inches in length. As in other chimaeras, the jaws are strongly calcified but have a rough, porous surface that does not look quite like bone.

Edaphodon mirificus Leidy

Edaphodon mirificus Leidy
Edaphodon mirificus, Applegate, 1970, pp. 392–393, fig. 175A.

Edaphodon barberi Applegate (fig. 37)

Edaphodon barberi Applegate, 1970, p. 390, fig. 174.

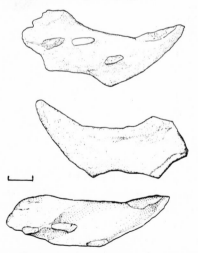

FIG. 37. *Edaphodon barberi* Applegate. Holotype mandible, FMNH PF 290. Top: medial view. Middle: lateral. Bottom: occlusal. Outlined areas are tritors. Scale bar 10 mm.

Edaphodon sp.

Edaphodon sp., Applegate, 1970, p. 393, fig. 175B.

We will depart here from our usual procedure and discuss these species together. The taxonomy of *Edaphodon* (and of the other chimaeras) is based on the shape of the jaws and on the number, position, size, and shape of areas of crushing teeth on the jaws. These areas of tiny teeth are referred to as *tritors*. There are four or five of these on an *Edaphodon* mandible.

We reproduce Applegate's figures, reducing them to the same scale. Line drawings show the outline of the tritors. Careful comparison with these illustrations should help identify your specimen.

Chondrichthyes *incertae sedis*

Order Bradyodonti

Incertae sedis literally means "of uncertain position." After well over a century of effort, we are little closer to a settlement of the relationships of these forms than were our predecessors. This "order" is a very heterogeneous assemblage, sharing only the commonality of being late Paleozoic shell-crushers with cartilaginous skeletons. The ancestors of the chimaeras lie probably within this group; exactly where they belong is a matter of dispute.

FAMILY PETALODONTIDAE

?Petalodus sp.

We have seen only a single specimen of this fish, collected from Mississippian rocks near Huntsville. As we had no opportunity for a detailed examination, and the specimen was in a private collection, its affinities remain uncertain. However, it did show the characteristic features of the genus.

The tooth is not unlike a fresh rose petal in shape. It shows the same double curve, giving an S-shaped lengthwise section. The base and crown both are spoon-shaped—the base concave downward and the crown concave upward.

FAMILY PSAMMODONTIDAE

Psammodus sp.

This species was reported first from Alabama in the pioneer work of Tuomey (1850B, p. 69) and noted again in his second report (Tuomey, 1858A, p. 39). His specimens came from Mississippian limestones near Huntsville and Florence.

The teeth are broad, crushing plates with a pitted surface. The name means "sand tooth," referring to the pitting. Specimens from limestones commonly have a bluish color because of leaching by ground water. Although Tuomey's specimens were war casualties, his description rings true, so there is little doubt that the identification was accurate—they can be "distinguished by their bluish color and punctated surface" (Tuomey, 1850B, p. 69).

Fragments are common in the Bangor Limestone (Upper Mississippian) in some areas. Entire teeth, up to 8 centimeters across, exist in private collections.

Class Osteichthyes
bony fishes

These abundant and varied fishes are distinguished from the cartilaginous fishes by the presence of true bone in the skeleton. It was once thought that the cartilaginous fishes were primitive and ancestral to the bony fishes. This opinion is no longer held. True bone was widely present in even more primitive vertebrates such as the Agnatha and the Placodermi. The bony and cartilaginous fishes now are viewed as independent, parallel developments from the placoderm stock.

Technically, the bony fishes are distinguished from the placoderms by the structure of the jaw, along with the gill arches from which the jaw is ultimately derived. Jaws arose among the vertebrates as modifications of the first (perhaps the second) original gill arch. This degree of specialization is as far as matters go among the placoderms; the other arches are unspecialized, and the jaw is supported only by direct contact with the skull. In the Osteichthyes, the next gill arch (hyoid arch) becomes modified into a jaw support. Particularly important is the most dorsal element of the hyoid arch, termed the hyomandibular in the bony fishes. This stout prop braces the jaw joint against the back of the skull. A similar development takes place in the Chondrichthyes.

Subclass Sarcopterygii
lobe-finned fishes

This subclass is not known in the Alabama fossil record. It is important mainly as a group that contains the beginnings of the land vertebrates. They may yet be found in the Paleozoic rocks of Alabama. The lobe-fins range from Devonian to Recent.

Subclass Actinopterygii
ray-finned fishes

The actinopterygians are the most common and varied of living fishes, and their fossil record is vast. A distinguishing feature is the fin, which is supported by slender rays rather than by a fleshy lobe. If a lobe is present, it is confined to the very base of the fin.

Most work that has been done on ray-finned fishes has been based on complete skeletons, found in only a comparatively few places. A more recent development is the identification of bony fishes on the basis of single bones. Most of our knowledge of Alabama bony fishes is based on such remains. Only a few complete fishes have been found as fossils in Alabama, mostly in the Mooreville Chalk.

A still more recent development is the use of single scales in identification. Such scales can be found in many places that do not yield numerous bones. This study is still in its infancy, and we must largely ignore it here.

Again, this subclass is known first in the Devonian. It is divided into three infraclasses.

Infraclass Chondrostei

This is a very primitive group of ray-finned fishes. The tail is heterocercal, like that of a shark. In fossil forms the body often is covered by a mail of thick, enameled (ganoid) scales, but scales are all but absent in living forms. The main bodies (centra) of the vertebrae remain cartilaginous throughout life.

The only living chondrosteans are the sturgeons, found worldwide, and the strange paddlefishes, found only in the United States and China. Sturgeons are very susceptible to pollution and are thus becoming rare. Paddlefishes, on the other hand, seem to be comparatively pollution-tolerant and may be increasing in numbers as their competitors are killed off. Unfortunately, paddlefishes are of little economic use, while the sturgeons are prized as food in many areas and their eggs form the finest and most expensive caviar. Sturgeons probably are also the largest living freshwater fishes. In former times, 1200-pounders were not particularly uncommon, and specimens weighing over a ton were known. Such large sturgeons are now exceedingly rare because of overfishing and pollution. The reason for overfishing is obvious, as a 1200-pound female may well yield over a thousand dollars in caviar.

A fossil sturgeon is known from the Cretaceous of Alabama.

Order Acipenseriformes
sturgeons

FAMILY ACIPENSERIDAE

Living sturgeons have very little bone in the body and are rare as fossils; however, earlier forms were much better ossified. Sturgeons are known definitely from the Jurassic and may date as far back as the Pennsylvanian (Romer, 1966).

Propenser hewletti Applegate (fig. 38)

Propenser hewletti Applegate, 1970, pp. 399–401, figs. 180–185.

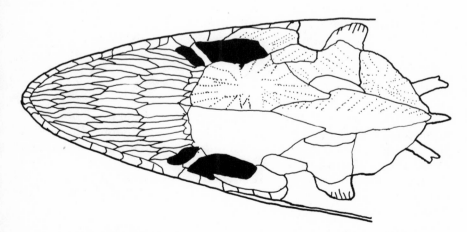

FIG. 38. *Propenser hewletti* Applegate. Restored skull, dorsal view. Ornamentation indicated on right side only. After Applegate.

Two specimens of this form, one a nearly complete skeleton, are known from the Mooreville Chalk of Greene County. Both are in the collections of the Geological Survey of Alabama.

Applegate (1970) gives a very thorough description of these specimens. The most distinctive character is the ornamentation of the skull plates, which are covered with hemispherical tubercules about 5 millimeters across, usually arranged in rows. For further information about other bones, including fin spines, dermal scutes, and vertebrae, see Applegate (1970).

Infraclass Holostei

The holosteans are intermediate ray-finned fishes. The vertebrae are completely ossified in most forms. They retain the ganoid scales and heterocercal tails of the chondrosteans. A few forms have lost most of the scales, and others have tails that almost obtain a homocercal shape, as in the teleosts. The oldest holosteans are from the Permian. Only two groups are living, both in North America. These are the gars *(Lepisosteus)* and the bowfin or grinnel *(Amia calva)*.

Order Semionotiformes

Suborder Semionotoidei

Most of these extinct primitive holosteans have shell-crushing dentitions. The body is armored with extremely massive ganoid scales.

FAMILY HADRODONTIDAE, *familia nova*

This family is based on a single genus known from the Upper Cretaceous of Mississippi, Alabama, and Kansas. This genus, *Hadrodus*, has been associated with the pycnodonts by most authors (*e.g.*, Applegate, 1970). Thurmond (in press) considers it allied to the semionotoids and has erected a new family to contain it.

Hadrodus priscus Leidy (fig. 39)

Hadradus priscus Leidy, 1857F, p. 167.
Hadrodus priscus, Applegate, 1970, p. 401, fig. 187.
Pycnodontidae . . . *incertae sedis* Applegate, 1970, pp. 401–403, fig. 187.

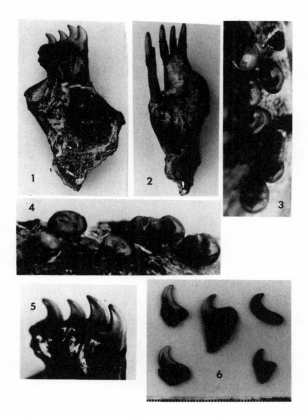

FIG. 39.   *Hadrodus priscus* Leidy. 1, 2, 3: pharyngobranchial(?) in medial, anterior
views and detail of medial surface. 4, 5: right dentary, detail of teeth in
occlusal and medial views. 6: isolated teeth from pharyngeals. Greene
County, Alabama, Mooreville Chalk. 1, 2, 6: x1; 3, 4, 5: x2.

Three portions of this remarkable fish are known thus far. Leidy
(1857F) described under this name a strange premaxilla with two incisorlike teeth from the Cretaceous of Mississippi. The incisors are
flattened blades about ½ inch across, with a median notch in each
tooth. Applegate (1970) added a lower jaw to our knowledge of this
species. This jaw is from Greene County and bears "crushing" teeth
in two rows. Thurmond (in press) analyzes this dentition and notes
that it is far from simple. The pharyngeals (tooth-bearing gill arch
elements) are the most distinctive bones. These bear large bladelike
teeth, flattened from side to side rather than from front to back as is
normal. Each tooth bears a large, hook-shaped cusp at its anterior
edge.

Suborder Lepisostoidei
gars

The relationships of this group to the semionotoids are disputed. Lehman (*in* Piveteau, 1966) considers them a separate order, Lepisosteiformes. We follow Romer (1966).

The gars are familiar to almost every Southern fisherman. They are elongated fishes with massive scaly coverings and long jaws well armed with sharp teeth. Usually considered inedible, their economic influence is largely negative. They are highly predaceous and often eat more desirable fishes. Many an expensive fish-stocking program has ended up as gar food. Scales were used occasionally by Indians as ornaments or projectile points.

There is only a single family, Lepisosteidae, known from the Lower Cretaceous.

FAMILY LEPISOSTEIDAE

In addition to the characters mentioned above, fossil lepisosteids may be recognized by their teeth and vertebrae. The teeth are slender, extremely sharp cones with flaring bases. The bases are fluted, and enamel is confined to a small cap at the point. The vertebrae are elongated and opisthocoelous (each vertebra having a socket at the back end and a corresponding bulge at the front), unlike those of any other fishes.

Fossil lepisosteids are known from the Eocene of Alabama. We have collected material from the Tallahatta Formation in Monroe County and from the Gosport at Little Stave Creek, Clarke County. Teeth only are known from the Gosport, but the Tallahatta has produced a partial lower jaw and probable skull plates. The jaw is indistinguishable from that of the living spotted gar, *Lepisosteus oculatus*. If the skull plates recovered belong to the same animal, however, they represent a distinct gar that had skull bones ornamented with denticles of ganoine rather than beads as in living forms.

Gars, as mentioned above, have thick, heavy scales. As yet such scales have not been recovered as fossils in Alabama.

Order Pycnodontiformes
pycnodonts

The pycnodonts are an odd-looking group of extinct, shell-crushing holosteans. The body is very deep, often almost circular, and considerably flattened from side to side. The scales are extremely modified. Instead of the complete covering of heavy ganoid scales common among the holosteans, scales are found only along the back ridge and the belly. The ridge scales have saddlelike bases and one or more denticles in a row. The scales lateral to these ridges consist of small, diamond-shaped plates, each with a long rod growing out of one corner. These rods interlace to form a basket-shaped mesh of unknown function around the entire body.

The most common pycnodont fossils are teeth and jaws. There are several parts to the jaws. Anteriorly, on the dentary in the lower jaw and the premaxillae in the upper are incisorlike teeth. The splenials in the lower jaw and the vomer in the upper bear a series of flattened crushing teeth. The vomer is along the midline of the body and is thus bilaterally symmetrical; the splenials meet at the midline, so that each bone is asymmetrical.

Pycnodonts are known from Upper Triassic to Eocene. They have been found in Alabama only in the Cretaceous.

FAMILY PYCNODONTIDAE

Anomoeodus sp. (fig. 40)

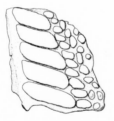

FIG. 40.   *Anomoeodus* sp., right splenial, AUMP 1695, Harrell Station, Dallas County, Alabama. *c.* x1.

This pycnodont has not been reported previously from Alabama, nor has any other except for referrals of *Hadrodus* to this family. Our occurrence is based on a partial splenial from the Tombigbee Sand Member of the Eutaw Formation from Perry County, an isolated tooth from the Mooreville Chalk in Hale County, and a fine splenial from Harrell Station, Dallas County (fig. 40). The splenial teeth of the largest series are very distinctive in *Anomoeodus*. They are much wider than long and arranged in a slanted row *(en échelon)* on the splenial. In occlusal view, they are often comma-shaped or faintly sigmoid.

As *Hadrodus*, referred to the pycnodonts by Applegate (1970), is now believed to be a semionotoid, these are the only known occurrences of pycnodonts in Alabama. The specimens are in the collection of the Geological Survey of Alabama and the Auburn University Museum of Paleontology.

Order Amiiformes
bowfin and relatives

The only living representative of this order is the bowfin or grinnel *(Amia calva)* of the southern United States. The bowfin is a very advanced holostean fish. The tail is barely heterocercal; indeed, at first glance it looks fully homocercal. The scales are very thin, as in the teleosts, but retain traces of an enamel covering.

Most of the Amiiformes seem to have been highly predaceous fishes, as *Amia* is today. The living bowfin is a past master of the art of the slow, silent stalk and the blindingly fast strike that covers the last few inches to the prey. Its dorsal fin and tail are its secrets. Slow rippling of the dorsal fin moves the fish toward its victim without any disturbance in the water. The broad, unforked tail provides the push for the strike, driven by the powerful back muscles. The Amiiformes are known from the Triassic to the Recent.

FAMILY PACHYCORMIDAE

This family comprises an extinct group of fast-swimming fishes with the front of the skull elongated and solidified into a bony rostrum. They are known from the Upper Triassic through the Cretaceous. Alabama representatives are known only from the Mooreville Chalk.

Protosphyraena nitida? (Cope) (fig. 41)

?*Erisichthe nitida* Cope, 1872EE, p. 280.
?*Erisichthe nitida*, Cope, 1875E, pp. 217, 275, pl. 48, figs. 3–8.
Protosphyraena nitida?, Applegate, 1970, p. 404.

The name of the genus comes from a time when the fish was thought to be a possible ancestor of the barracudas *(Sphyraena)*. Although this opinion is no longer held, the name, according to the rules, has stuck.

There are two instantly recognizable bones in any *Protosphyraena*: the rostrum and the pectoral fin spine (see fig. 41). The rostrum is a long, stout, slightly curved bone. There are weak flutings along its sides, and its base has one or more slender, daggerlike teeth that are

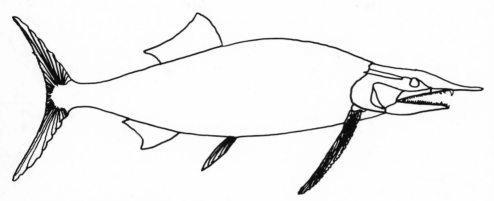

FIG. 41. *Protosphyraena*, restored outline. The rostrum and the large, saw-edged, pectoral fin spines are easily recognized. Length about two meters. After Gregory.

sloped forward. We interpret that *Protosphyraena* stunned or gashed its prey with the rostrum before engulfing it. The fin spine is unmistakable. From a faceted base for articulation with the bones of the body projects a huge, flattened, scimitar-shaped blade of bone, sometimes well over 2 feet long.

Applegate (1970) notes the possible presence of another species of *Protosphyraena* in the Mooreville Chalk, based on differences in fin spine ornamentation.

Infraclass Teleostei

The teleosts form by far the most abundant living group of fishes. The preceeding infraclasses, Chondrostei and Holostei, survive as bare remnants among the vast numbers and variety of living teleosts. The teleosts appeared in the Early Jurassic. By the Late Cretaceous, they had achieved dominance over the older groups and increased their sway ever since.

The scales invariably are thin and show no enamellike layer. Essentially, they are very thin plates of bone, overlapping like the shingles of a roof. There are two basic types of teleost scales. Cycloid scales are simple bony plates, with growth lines on the exposed portions. They are found in most primitive teleosts. Ctenoid scales have rows of small denticles on the exposed portions; they are especially characteristic of the perchlike fishes, very advanced teleosts. A few teleosts, like the eels and many catfishes, have lost their scales completely.

It is important to understand the scale structure of teleosts. Each scale is built around a center of growth, the nucleus. The shape, position, and ornamentation of the nucleus may be of significance in identifying the scale. Around the nucleus are concentric rings of growth (like tree rings), the circuli. These are not annual rings; rather, each ring marks a pause in the growth of the scale. In some cases, annual interpretations may be placed on the relative widths of circuli, for they are narrow during periods of slow growth and wide during rapid growth. Ridges radiating out from the nucleus, the radii, are present on the covered portion of the scale in most cases. Many identifications of scales are based on the form of the circuli and the number and arrangement of radii.

The structure of the fins and skull (including jaws) is extremely variable in the teleosts. Teeth are common as fossils. Important characters to note are the shape and means of implantation in the jaw.

Order Elopiformes

These are primitive teleosts with cycloid scales. The living representatives include the ladyfishes, or skipjacks, and the bonefishes. There are numerous extinct groups.

Suborder Elopoidei

The skipjacks and their relatives in this suborder represent a primitive stock of teleosts off the main line of development, which leads to the herrings, trouts, and others.

FAMILY ELOPIDAE
ladyfishes or skipjacks, tarpons

In general, two quick tests help to recognize a fish as a primitive or an advanced teleost. The first is to examine the dorsal fin. Is it composed of comparatively soft and flexible rays more or less alike distributed throughout the fin, or does it have the anterior rays hardened into heavy spines? The first type is primitive, as in the Elopiformes; the second is extremely advanced, found in the perches and their relatives.

The second step is to check the position of the pelvic fins, the posterior paired fins on the belly. If they are posterior, usually near the anus, the fish is primitive. On the other hand, pelvic fins far forward, ventral to the pectoral fins, indicate an advanced fish. The elopoids are a very conservative group retaining both primitive characters. They are normally inedible, as the development of bones between the muscles makes a prickly mouthful. However, they are scrappy fighters on a fishing line! One member of the order, the tarpon *(Megalops)*, is among the world's most sought after game fishes.

Palelops eutawensis Applegate (fig. 42)

Palelops eutawensis Applegate, 1970, p. 406, fig. 188.

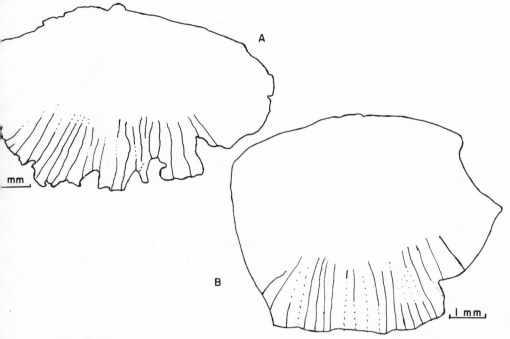

FIG. 42.   *Palelops eutawensis* Applegate. Scales. A: FMNH PF 3559, type. B: FMNH PF 3560, schematic drawing shows radii (dotted where obscured). Circuli not shown; mostly obscured by granulate ornamentation covering scales. After Applegate. x10.

This species (and the genus based on it) is known only from scales. The original specimens come from Alabama, 6.2 miles west of Aliceville, Pickens County. Others have been found in Greene County. All of the Alabama material comes from the Mooreville Chalk. The name is based on a misconception, as Applegate (1970) refers to the locality as being in "Eutaw County," which does not exist, although Eutaw is a town in Greene County. Similar scales are known from Upper Cretaceous rocks in California, Kansas, and South Dakota.

The scales are simple and either oval or subrectangular. The nucleus and much of the rest of the scale are covered with a granular ornamentation. The circuli are straight on the dorsal and ventral portions of the scale, and the nineteen to twenty-one radii are more or less parallel. Applegate (1970) states that this genus differs from the living *Elops*, the skipjack, in having more radii, ornamentation extending over the nucleus, and granular ornamentation on the apicial (exposed) portion of the scale. This ornamentation is vermicular (literally "wormy") in *Elops*.

FAMILY PACHYRHIZODONTIDAE

The name of this group literally means "with thick-rooted teeth." The teeth are stout, somewhat blunt cones in most forms, with widely expanded bases deeply set in sockets. A jaw of *Pachyrhizodus* looks more like a miniature mosasaur than a fish; however, the smooth, shiny, almost enamellike surface quickly betrays its fishy origin. These jaws attain a length of 1 foot or more in some species. *Pachyrhizodus* and its near relatives are found only in Cretaceous rocks but are known over most of the world.

Pachyrhizodus minimus Stewart (fig. 43, 44)

Pachyrhizodus minimus Stewart, 1899B, p. 37.
Pachyrhizodus minimus, Stewart, 1900A, p. 361, text-fig. 1; Applegate, 1970, pp. 408–410, figs. 189, 190A, 191B–C, 192.

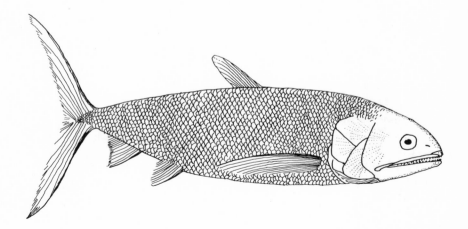

FIG. 43.   Reconstruction of *Pachyrhizodus minimus* Stewart. Length about 1 meter. After Applegate, 1970.

This species of *Pachyrhizodus* is small (about 3 feet long). It is known in Alabama from a nearly complete fish from Greene County and a few fragments from Dallas County (Applegate, 1970). The teeth are proportionately very small and slender. There are very fine longitudinal striae on the sides of the vertebrae, and cross sections of vertebrae are not complex, showing an outer wall with about eight simple supporting struts in the central cavity.

The scales of this species have about seventeen radii in a triangular region on the exposed portion. The circuli are comparatively few and coarse.

Pachyrhizodus caninus Cope (fig. 44)

Pachyrhizodus caninus Cope, 1872I, p. 344.
Pachyrhizodus leptopsis Cope, 1874C, p. 42.
Pachyrhizodus caninus, Cope, 1875E, pp. 221, 276, pl. 50, figs. 1–4; Applegate, 1970, pp. 410–411, figs. 190B–C, 191A, 193, 199B.

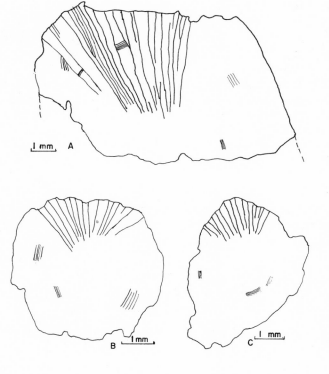

FIG. 44.   Scales of Alabama species of *Pachyrhizodus*. Schematic drawings showing radii and representative parts of circuli. A: *P. caninus*. B, C: *P. minimus*. Note fewer, more even radii of latter. After Applegate, 1970.

A giant among the Mooreville fishes, *Pachyrhizodus caninus* has jaws that attain a length of 1 foot, representing about 8 feet of fish. The teeth are stout, slightly curved cones, rising from flaring, weakly fluted bases set in definite sockets. The sides of the vertebrae are smooth. Vertebral cross sections show an extremely complex structure, with numerous supporting slats along the outer walls and separate supporting columns in the interior. Scales have finer and more numerous circuli than those of *P. minimus*.

Pachyrhizodus kingi Cope

Pachyrhizodus kingi Cope, 1872I, p. 346.
Pachyrhizodus kingi, Cope, 1875E, pp. 223, 276, pl. 46, fig. 11; Applegate, 1970, p. 411, fig. 190D.
Pachyrhizodus leptognathus Stewart, 1898D, p. 193, pl. 17, fig. 1.
Pachyrhizodus velox Stewart, 1898D, p. 193, pl. 17, fig. 2.

Although smaller, *Pachyrhizodus kingi* is much like *P. caninus*. Applegate (1970) considers that it might be the juvenile form of *P. caninus* but separates it because intermediate sizes are not known. The simplest distinction can be based on the vertebral cross section, which shows comparatively simple V-shaped supporting struts with a few separate columns.

All the species of *Pachyrhizodus* known from Alabama originally were discovered in Kansas. The Alabama forms are from the Mooreville Chalk.

Suborder Albuloidei

This group comprises the bonefishes and their relatives, all slender, fast-swimming forms. Their diet is principally shellfish, and they have powerful crushing teeth in a pavement on the palate, parts of the lower jaw, and on the gill arches.

FAMILY ALBULIDAE
bonefishes

The bonefishes are among the most avidly pursued game fishes by the light-tackle angler. They are noted for their wily ways, their determined battles, and, above all, their blinding speed.

There are many technical details that help to differentiate the bonefishes. A primitive feature is the presence of a gular plate. This bony plate fills the space between the bones on the ventral surface of the lower jaw. In most teleosts, this space is filled with thin slats of bone, called branchiostegal rays. The maxilla, in primitive fishes the major external bone of the upper jaw, no longer contacts the margin of the mouth (an advanced feature). The jaws, palate, and gill arches are covered by a mosaic of fine conical teeth. The parasphenoid (bottom of the braincase) bears a covering of large button-shaped teeth, as does the basibranchial (ventral element of the gill arches).

The earliest bonefishes are known from Lower Cretaceous rocks.

Albula dunklei Applegate (fig. 45)

Albula dunklei Applegate, 1970, pp. 412–413, figs. 194, 200B–C, 201B.

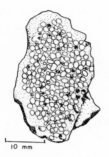

FIG. 45. *Albula dunklei* Applegate. Basisphenoid tooth plate, showing tooth arrangement. After Applegate. x1.

This is an ancient representative of the Recent bonefish, *Albula vulpes*. It is known only from very fragmentary remains, most of which could be distinguished most readily by direct comparison. These remains have been found in the Mooreville Chalk over a wide area in west-central Alabama.

The most distinctive features of the species are the parasphenoid bone, which is covered with minute crushing teeth (about 1/50 inch or .5 millimeter in size) of irregular shape, and the scales. Albulid scales are very distinctive. They have no true radii, but the basal portion of the scale shows strong folds that are carried out into scallops along the basal margin. There are four such scallops in *A. dunklei*. The distinct, fine, closely spaced circuli on the dorsal and ventral parts of the scale break up into vermiculate (wormlike) ridges on the basal portion. The basal portion is ornamented with rows of minute granules in *A. dunklei*; the granules of *A. vulpes* are in a dendritic (treelike) pattern.

Albula sp.

Applegate (1970) notes the possibility of another species of *Albula* from the Mooreville Chalk of Greene County. It is known only from a fragment of palate, bearing teeth twice the size of those of *A. dunklei*.

FAMILY BANANOGMIIDAE

An extinct family, similar to the bonefishes, the bananogmiids are known only from the Cretaceous. They differ from the albulids in several details. There is no gular plate, and the maxillae form part of the border of the upper jaw. The scales show distinct circuli on the basal as well as the dorsal and ventral portions. Scale radii are almost parallel to each other, arranged anteroposteriorly, numerous and fine but distinct. The granular ornament of the basal portion is arranged in rows parallel to the radii.

Bananogmius crieleyi Applegate

Bananogmius crieleyi Applegate, 1970, pp. 414–416, fig. 196.

This species is another described from miscellaneous fragments. Only one occurrence is known, from the middle part of the Mooreville Chalk of Dallas County, Alabama. The most distinctive bone is the parasphenoid (a palatal bone). This long, narrow slat of bone bears a large, club-shaped patch of teeth. In general, the skull bones are much larger and more massive than those of *Albula*. See family description for discussion of the scales.

Applegate (1970) notes the possible presence of up to three more species of *Bananogmius* from the Mooreville Chalk of Alabama. The remains are too fragmentary to be discussed, or illustrated, here.

Moorevillia hardi Applegate

Moorevillia hardi Applegate, 1970, p. 416, fig. 197.

This form is very similar to *Bananogmius*. The most distinctive bone of *Moorevillia hardi* is the premaxilla; this is proportionately smaller than the premaxilla of *Bananogmius,* and its external surface is not ornamented. In *Bananogmius* this bone is covered externally with large pits, much like the tooth sockets of the jaw elements.

This species is known only from a single specimen from the middle part of the Mooreville Chalk of Dallas County, Alabama.

Order Anguilliformes
eels

Two species of conger eels have been described from the Tertiary of Alabama, based on otoliths (ear bones). As the study of otoliths is a highly abstruse subject, and their collection not the usual province of the amateur, we will only note names and papers to see for further information.

Conger brevior (Koken)
Conger sanctus Frizzell and Lamber

See Koken (1888A, p. 293, pl. 18, fig. 7) and Frizzell and Lamber (1962).

Order Clupeiformes
herrings and relatives

It may be best to note at this point that the classification of the teleostean fishes is the subject of considerable controversy. Each specialist seems to have a pet scheme, and all classifications are subject to perpetual modification. For an idea of the confusion that has resulted, anyone who wishes to pursue the problem further might see the classifications of Romer (1945, 1966) and Berg (1947), among others. Such prestigious works as Lehman (*in* Piveteau, 1966) have appeared to skirt the real questions that exist by treating the bony fishes on a faunal rather than a taxonomic basis.

We should like to exercise a similar degree of discretion here, but the format of this work requires that we adopt a classification and stick to it. We will use that of Romer (1966). It is not particularly satisfactory; in general, it is far too cumbersome and complex. It has the virtue of being thorough, especially on the fossil forms, and is fairly readily available. A disadvantage is that Applegate (1970), the best work available on Alabama Cretaceous fishes, strongly disagrees with Romer's classification.

The Clupeiformes as viewed by Romer (1966) are a homogenous group of small, primitive teleosts, very near the ancestry of all more advanced forms but without the specializations of the Elopiformes. In particular, bones between the muscles are much less common in the Clupeiformes. For this reason, many members of the order are edible and constitute one of the larger parts of the world's fish catch (herrings, sardines). They draw very little attention from sport fishermen.

FAMILY CLUPEIDAE
herrings

Indeterminate herring remains are reported from the Mooreville Chalk by Applegate (1970). He notes three specimens: a preopercular bone (part of the gill cover) and two scales. While he does not identify the preoperculum further, he considers the scales to belong to the subfamily Dussumieriinae (the round herrings, placed as a separate family Dussumieriidae by some workers). These scales are certainly distinctive, even though incomplete. Circuli are very fine and indistinct. There are four very strong radii, with widely spaced, pronounced ridges at right angles to the radii.

Order Osteoglossiformes

The only living representatives of this order are some peculiar South American freshwater fishes. The largest and best known is the pirarucu *(Osteoglossum)*, a native of the Amazon Basin, weighing up to several hundred pounds. The fossil representatives of the order are numerous, especially in Cretaceous rocks. Most of them are marine, and many are large. The teeth are sharp, fairly slender cones, planted firmly in deep sockets.

Romer (1966) includes the Bananogmiidae (under the name Plethodontidae or Thryptodontidae) in this order. This is one of the few places we have followed Applegate (1970). See above for discussion of Applegate and Romer.

FAMILY ICHTHYODECTIDAE

These fishes make up one of the most conspicuous elements of the fauna of the Mooreville Chalk. They are large fishes, so their abundance relative to other fishes is difficult to estimate—large specimens are more likely to be seen and collected than are small ones.

The most distinctive bone is the lower jaw. This is a deep, massive bone bearing large teeth firmly planted in sockets. There are no foramina (openings) for blood vessels along the inner margin of the jaw at the tooth bases. The teeth are rounded in cross section.

Ichthyodectes sp. cf. *I. ctenodon* Cope

?*Ichthyodectes ctenodon* Cope, 1870L, p. 536.
?*Ichthyodectes ctenodon* Cope, 1875E, pp. 207, 241, 274, pl. 46, figs. 1–4; Stewart, 1900, pl. 49, fig. 5.
Ichthyodectes cf. *ctenodon*, Applegate, 1970, p. 418.

This species is best known from the Niobrara Chalk (Upper Cretaceous) of Kansas. Only a single jaw fragment is described thus far (1980) from Alabama, from the middle of the Mooreville Chalk of Greene County.

For the basic features of the lower jaw, see above. In *Ichthyodectes* the anterior teeth of the jaws are similar in size to the other teeth. In the closely related *Xiphactinus* these anterior teeth are enlarged into powerful fangs.

Xiphactinus audax Leidy (fig. 46)

Xiphactinus audax Leidy, 1870H, p. 12.
Xiphactinus audax, Leidy, 1873B, pp. 290, 348, pl. 17, figs. 9–10; Cope, 1875E, p. 276; Applegate, 1970, p. 418, fig. 200A.

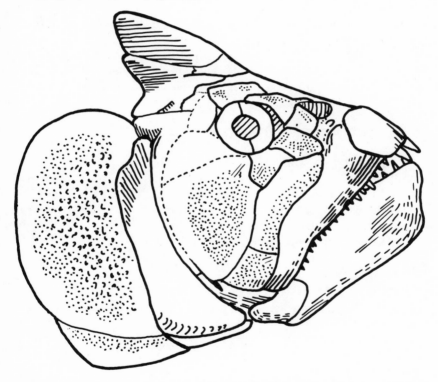

FIG. 46. *Xiphactinus audax* Leidy. Reconstructed skull from Kansas. Redrawn from Bardack, 1965, fig. 4. x0.2.

Of the common Cretaceous bony fishes, *Xiphactinus audax* is the most spectacular. Specimens up to 12 or 15 feet long are known from Kansas. It is well known enough to have a more-or-less popular name or two, either "bulldog fish" or "Cretaceous giant tarpon." Many past and present authors refer remains of this fish to Cope's name *Portheus molossus* because of doubt about the validity of Leidy's species, based only on a single pectoral fin spine (the name can be translated as the "bold sword-spine"). Cope's species is based on a nearly complete skull that is magnificently preserved.

The lower jaw is similar to that described for the Ichthyodectidae in general, except for the enlargement of the anterior teeth into powerful fangs, often 2 inches or more long. The vertebrae are probably the most common finds. As with most teleosts, the vertebral centra are roughly disk-shaped, with a deep conical pit in each end. These pits often join through a small hole in the center that contained the notochord. The sides of the vertebrae bear extremely deep longitudinal furrows or pits. There are three of these on each side, a large but shallow one dorsally, followed by a narrow but deep groove, and then by a very large and deep pit.

It is doubtful that all vertebrae of *Xiphactinus* can be distinguished reliably from those of *Ichthyodectes*. The latter are smaller, usually less than 2 inches across, and the lateral furrows are more even in size.

FAMILY SAUROCEPHALIDAE (SAURODONTIDAE)

This Upper Cretaceous family can be distinguished from the Ichthyodectidae by two characters. First, the teeth are compressed and bladelike, rather than rounded. Second, the lingual surface of the lower jaw bears a series of notches or foramina (holes) for blood vessels and/or nerves at the tooth bases. The lower jaw has a predentary bone (in front of the main tooth-bearing bone, the dentary) that is lacking in the Ichthyodectidae.

Saurodon leanus Hays

Saurodon leanus Hays, 1830, p.477, pl. 16, figs. 1–10; [*fide* Hay, 1902A, p. 386. Cope, 1875A, gives p. 476].
Saurocephalus leanus, Hay, 1902A, p. 386; 1929, p. 740.
Saurodon leanus, Applegate, 1970, pp. 419–420, fig. 200E.

The jaw of this species has deep notches, rather than foramina, on the lingual surface below the alveolar border. The vertebrae are superficially similar to those of *Xiphactinus* but smaller and with only two longitudinal furrows rather than three.

This fish was first described from the Upper Cretaceous of New Jersey. It remains are very abundant in the Mooreville Chalk of Alabama to judge from Applegate (1970) who notes seventeen identifiable specimens. Applegate (1970) notes the possible presence of two other species from the Mooreville Chalk.

Saurocephalus sp. cf. *S. lanciformis* Harlan

?*Saurocephalus lanciformis* Harlan, 1824A, p. 331, pl. 12.
Saurocephalus cf. *lanciformis*, Applegate, 1970, p. 420.

This was one of the first fossil fishes to be described from North America. The original specimen (Harlan, 1824A) was collected in 1804 by the Lewis and Clark Expedition near what is now Council Bluffs, Iowa. The original description considered the specimens to represent a reptile related to the plesiosaurs, but the fishy affinities of the beast long have been known.

The lower jaw differs from that of *Saurodon* in having foramina for blood vessels rather than notches at the bases of the teeth on the inner surface of the bone.

Order Salmoniformes
salmon, trout, etc.

Living representatives of this order are a large and important group in both fresh and marine waters. Perhaps their most easily recognized feature is the presence of an adipose fin, a small fleshy dorsal fin posterior to the main dorsal. It is also present in the catfishes and in certain minnowlike fishes.

FAMILY ENCHODONTIDAE

This extinct family of salmonlike fishes was very important in Cretaceous and Early Tertiary times. The jaws bear two rows of teeth, with the labial row much smaller than the lingual. On the palatine bones (in the roof of the mouth) are one or two enormous fangs, often shaped like a stilletto. In the genus *Eurypholis*, these teeth are barbed and might be confused with those of scabbardfishes (Trichiuridae) or some early sawfishes (Pristidae, subfamily Ganopristinae). This genus has not been found yet (1980) in Alabama.

The lower jaws have a unique symphysis (junction between the two lower jaws). Both jaw halves send out long, rounded processes that fit into corresponding notches in the other side, not unlike the interlocking of pieces in a jigsaw puzzle.

Enchodus petrosus Cope

Enchodus petrosus Cope, 1874C, p. 44.
Enchodus petrosus, Applegate, 1970, p. 421.

The palatal fang of this species is long and slender. It is highly compressed with anterior and posterior carinae. The lower jaws are not highly ornamented.

Enchodus sp. cf. E. saevus Hay

?Enchodus saevus Hay, 1903, pp. 76–80, text-figs. 58–65.
Enchodus cf. saevus, Applegate, 1970, p. 421.

This species differs from E. petrosus in the ornamentation of the lower jaw and perhaps in the form of the palatal fang. The lower jaws bear radiating ridges over the outer surface. As described by Hay (1903), the palatal fang is almost D-shaped in cross section. The carinae (cutting edges) are on the lateral and medioanterior surfaces, with a flattened face between them on the anterolateral surface. Applegate (1970) considers this criterion unreliable, but it is useful at least as a rule of thumb on isolated palatines and fangs. This species is known from the Niobrara of Kansas and the Mooreville of Alabama.

Cimolichthys nepaholica (Cope)

Empo nepaholica Cope, 1872I, p. 347.
Empo nepaeloica, Cope, 1875E, pp. 230, 279, pl. 49, fig. 9, pl. 50, fig. 8, pl. 52, fig. 1, pl. 53, figs. 3–5.
Empo nepaholica, Hay, 1903A, p. 81, pl. 1, fig. 4, text-figs. 69–72.
Cimolichthys nepaholica, Hay, 1929, p. 756; Applegate, 1970, p. 421.

The teeth of this species are small, very slender, curved, and lean forward. The teeth have slightly flaring, weakly fluted bases, reminiscent of Pachyrhizodus on a miniature scale. However, as in the other enchodontids, they are fused to the jaws rather than being set in sockets.

This species was described (Cope, 1872I) from the Niobrara Chalk of Kansas. Applegate (1970) reports it from the Mooreville Chalk of Alabama, in Greene and Hale counties.

FAMILY DERCETIDAE

This strange Upper Cretaceous family of fishes has been of disputed relationships. We follow Applegate (1970) in placing it near the Enchodontidae. The jaws are covered with orderly rows of very slender teeth, especially on the palatines.

Stratodus apicialis Cope

Stratodus apicialis Cope, 1872I, p. 349.
Stratodus apicialis, Cope, 1875E, pp. 227, 279, pl. 49, figs. 6–8; Applegate, 1970, p. 421, fig. 198A–F.

The only dercetid known from Alabama, this fish is very distinctive. The numerous rows of teeth (or their sockets) on the jaws are diagnostic. As the teeth are very slender and delicate, they often are not preserved. Exceptions are the teeth on the premaxillae and maxillae; they are oddly shaped, with a cylindrical base and a flaring top capped by a blunt cone. *Stratodus* is known from the Niobrara Chalk of Kansas and the Mooreville Chalk of Greene and Dallas counties, Alabama.

Order Perciformes
perches and their relatives

The perches and their relatives, except for some specialized offshoots of the group, are the most advanced teleosts in modern seas. The dorsal fin is divided sharply into an anterior spiny portion and a posterior soft-rayed portion. The pelvic fins are far anterior, beneath and even in front of the pectoral fins. The pelvic girdle (the bony plates to which the pelvic fins are attached) is joined to the pectoral girdle. These fishes are doubtless the most abundant, varied, and successful forms today, both in fresh and salt waters.

Suborder Percoidei
perches, basses, drums, etc.

This suborder comprises the fishes that are generally considered "typical." In particular, it contains a very large proportion of the freshwater game fishes.

FAMILY SCIAENIDAE
drums

The drums are a specialized group of perchlike fishes with a dentition adapted for crushing shellfish. They are found in both fresh and salt waters. The name of the group comes from the booming noises made by many of its members.

Only a single drum tooth is known to exist at present in the fossil record of Alabama. This tooth is a recent discovery from the basal Gosport Sand at Little Stave Creek, Clarke County. The specimen is indistinguishable from teeth in the lower jaw of *Pogonias*, the sea drum, so identification by teeth alone is not precise. Drum jaws eventually should be found in the Eocene of Alabama.

Tuomey's (1850B, p. 162) record of *Lepidotus* (*Lepidotes*, a semionotoid holostean) from the Tertiary of Alabama is probably this genus, as the teeth are similar. The specimen presumably was a victim of General J. H. Wilson's raid on the University of Alabama in 1865.

Suborder Mugiloidei
mullets and barracudas

The two most prominent members of this suborder seem to be an ill-assorted pair at first glance. Mullets are small, inoffensive-looking fishes, mainly used as bait for other fishes. On the other hand, the barracudas are, pound for pound, the most feared fishes in modern oceans. All, however, are slender-bodied fishes with two distinct dorsal fins (the anterior spiny) and anteriorly placed pelvics. The latter feature is a quick way to distinguish the barracudas from the superficially similar pikes (family Esocidae), which have posterior pelvic fins.

FAMILY SPHYRAENIDAE
barracudas

The only virtue most humans are willing to concede to the barracudas is that they are smaller than sharks. Size at least keeps their menace within bounds. A barracuda is an ugly customer, with a torpedo-shaped body and an underslung lower jaw well armed with a potent set of daggers and slicers. The eye is very large ("the better to see you with, my dear"), as barracudas hunt mainly by sight and will attack almost anything that moves.

The earliest known barracudas are of Eocene age. Alabama material has been found in the Tallahatta and Gosport Formations (Claiborne Group, middle Eocene) and the Jackson Group (upper Eocene), in Monroe, Clarke, and Choctaw counties.

Sphyraena sp. (fig. 47)

Sphyraena cf. *major* Leidy, White, 1956, pp. 127, 147.

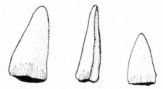

FIG. 47.   *Sphyraena* sp., Thurmond Collection 30, Little Stave Creek, Clarke County,
Alabama. Anterior tooth, lingual and anterior views; lateral tooth, buccal
view. x2.

Teeth of barracuda are fairly common in the Eocene of Alabama (see
above for details). Anterior teeth in both jaws are broad, sharp blades
with a slight S-curve in side view. The posterior teeth are smaller,
nearly straight, with two convex cutting edges. Both types of teeth are
found as fossils and represent the same fish. Until better preserved
material is available, it seems best to refer such remains to *Sphyraena*
sp., rather than attempting to speculate on their more precise relationships.

Suborder Labroidei
wrasses and parrotfishes

Today these fishes are inhabitants of warm seas, particularly of coral
reefs. They are principally shellfish eaters, and many species eat
mainly coral. The wrasses have large, stout, well-spaced nibbling
teeth, while the parrotfishes have front teeth that are fused together
into a parrotlike beak. Both have a massive crushing dentition on the
palate and gill arches.

Wrasses (Labridae) are known from the Upper Cretaceous rocks of
North America (but not in Alabama). The more specialized parrotfishes (Scaridae) appear in the Eocene.

FAMILY LABRIDAE

Phyllodus sp. (fig. 48)

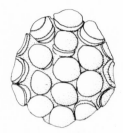

<small>FIG. 48.   *Phyllodus* sp., basisphenoid (upper) toothplate, BSC 781, near Georgiana, Butler County, Alabama. Note that replacement teeth show at edges of many teeth. Teeth are "stacked." x2.</small>

This species is reported herein for the first time from the Eocene of Alabama. The genus is characterized by its crushing toothplates. These consist of several teeth, usually a large series along the midline with a row of smaller teeth on each side and around the front. Each tooth is a thin plate, domed in the center, and underlain by its successor teeth, forming a pile not unlike the leaves of a book (hence the name of the genus, literally "leaf-tooth"). Until more complete toothplates are found, it is impossible to speculate further on their relationships. They seem particularly close to *Phyllodus toliapicus* from the London Clay of England (also Eocene). Isolated teeth are known from Alabama, in the Gosport Sand (middle Eocene) at Little Stave Creek, Clarke County.

Suborder Scombroidei
mackerels and relatives

This group consists of very fast-swimming fishes, all fairly slender-bodied. From this point, their structure is diverse. They include the very slender, flattened cutlassfishes or scabbardfishes (Trichiuridae), commonly used as bait in the Gulf of Mexico and called "ribbonfish" colloquially. The family Scombridae includes the familiar and tasty mackerels and tunas. The two families of billfishes, the swordfishes and the marlins and sailfishes, belong to this suborder. A characteristic feature of the group is the fusion of the last few vertebrae and their processes into a sturdy bony fan to support the tail. Such tail fans are among the most common fossils of the suborder but have not been reported definitely from Alabama.

Despite the lack of tail fans, the fossil record of this suborder is extensive in Alabama. Scombroids are abundant in the middle and upper Eocene rocks of Alabama. Isphording and Lamb (1971) also note the presence of an unidentified scombroid in the Pliocene of Mobile County.

FAMILY TRICHIURIDAE
cutlassfishes or scabbardfishes

The very slender, elongate fishes of this family have not been studied extensively in the fossil record, particularly in North America. One of the major surprises of the authors' collecting in southwestern Alabama was the presence of numerous teeth referable to the Trichiuridae.

Teeth of this family are unmistakable, as they are mostly slender, highly compressed blades, often strongly curved, with sharp cutting edges. The most striking feature is the presence of a barb, often powerful, at the tip.

Trichiurus sp. (fig. 49)

FIG. 49.   *Trichiurus* sp., BSC 782, tooth, near Peterman, Monroe County, Alabama. Note that barb is large, and enamel covers entire tip. x3.

Teeth referable here are fairly stout and large for the family, up to ¾ inch in length. The blade is bent sharply backwards near the base. The cutting edges are very prominent, but the barb is small. The preservation of the teeth is often peculiar, as the enamel commonly is almost completely removed except for traces along the edges. The anterior cutting edge is far more prominent than the posterior. Our Alabama specimens, from the Gosport Sand at Little Stave Creek, Clarke County, and from a locality near Burnt Corn, Monroe County, closely resemble those of *Trichiurus oshosunensis* White (1926A) from the middle and late Eocene of Africa. In particular, they resemble teeth of this species figured by Arambourg (1935F, p. 432, pl. xx, figs. 7–8). This species, or a closely related form, also appears in the Tallahatta Formation ¼ mile southeast of Peterman, Monroe County.

Trichiurides sp. (fig. 50)

FIG. 50.   *Trichiurides* sp., BSC 783, tooth, near Peterman, Monroe County, Alabama. Note that enamel only covers edges of the tip, and barb is minute. x3.

The teeth referred to here come from the Tallahatta Formation near Peterman, Monroe County, and the Gosport Sand at Little Stave Creek, Clarke County. They are minute, none over 5 millimeters (⅕ inch) in length. The tooth base is very slender and highly flattened. The barb is proportionately much larger than that of *Trichiurus,* often as wide as the base, and may extend beyond both edges of the base.

These teeth closely resemble those of *Trichiurides sagittidens* Winkler from the Paleocene of Belgium, as figured by Casier (1946, p. 133, pl. iii, fig. 14). As far as the authors can determine, this is the first record of the genus in North America.

Teeth of this genus are extremely difficult to collect, even by screening techniques. Numerous specimens were observed in material from Peterman during washing but seem to have passed through the screens. Further collecting and washing, using finer screens, is under way (1980).

?FAMILY BLOCHIIDAE
billfishes

Cylindracanthus rectus (Agassiz)

Coelorhynchus rectus Agassiz, 1843B, vol. 5, p. 92.
Cylindracanthus ornatus Leidy, 1856E, p. 12.
Coelorhynchus ornatus, Leidy, 1856E, p. 302; Cope, 1870F, p. 294; Woodward, 1891A, p. 122.
Cylindracanthus rectus, Fowler, 1911A, pp. 141–142, fig. 87; Leriche, 1942B, pp. 49–50, pl. 4, fig. 3; White, 1956, pp. 127, 147.

The name *Cylindracanthus* refers to elongate spines of circular or nearly circular cross section, slightly tapered, and bearing numerous flutes along the sides. These flutes are very straight and only occasionally branch toward the thicker end. As far as the authors know, no complete specimen has ever been found, especially no specimen showing any signs of contact with another bone. Yet material of this genus is quite abundant on both sides of the Atlantic.

Fowler (1911A) gives counts of the flutings on material from the Eocene of New Jersey ranging from 35 to 45. On two Alabama specimens, we count respectively 46 and 44. One of our specimens shows a deep groove on one face (?ventral). The interior is a large cavity, and the grooved specimen shows a thin septum (wall) of bone extending into this cavity immediately adjacent to the groove. The spines attain diameters of 20 millimeters (about ¾ inch) and lengths near 30 centimeters (1 foot).

The species is widespread in the upper Eocene (Jackson Group) of Alabama and Mississippi. It seems particularly abundant in road cuts in Clarke County. We have not yet collected it outside the upper Eocene in Alabama.

Cylindracanthus acus (Cope)

Coelorhynchus acus Cope, 1870F, p. 294.
Cylindracanthus acus, Hussakof, 1908A, p. 44, fig. 18; Fowler, 1911A, p. 142, fig. 88.

This species was described (Cope, 1870F) from the Eocene of New Jersey. It is doubtfully distinct from *Cylindracanthus rectus,* and many authors (*e.g.,* Leriche, 1942B) have considered it the juvenile form of *C. rectus.*

These spines of *C. acus* are much smaller than those of *C. rectus,* attaining a diameter of about 7 millimeters (¼ inch). The flutings are fewer in number (19 on the type, AMNH 2246) than in the larger species, and relatively coarser. None of these characters would prevent it from being considered a juvenile *C. rectus.* However, material of intermediate size is not known, and we find *C. acus* in Alabama in rocks older than the Jackson Group in which *C. rectus* is recovered. Our material consists of fragmentary spines from the Gosport Formation at Little Stave Creek, Clarke County, Alabama.

Order Tetraodontiformes
puffers, triggerfishes, etc.

This is an extremely varied group of small fishes, mostly of bizarre shapes. Many are shell-crushers, and some have a complete dermal armor. One representative, a Japanese puffer known as the fugu, is highly prized in Japan for food, though highly dangerous. It contains one of the most powerful poisons known, particularly concentrated in the gonads. Eating improperly prepared fugu may produce death within minutes. It would seem prudent to keep one's bill paid with a Japanese fishmonger.

FAMILY OSTRACIIDAE
trunkfishes

These fishes are noted for having the entire body encased in a rigid bony armor composed of a mosaic of small plates. Swimming is done entirely by the fins, rather than by body movements as in most fishes.

Ostracion sp.

White (1956, pp. 146–147) notes, but does not describe or figure, a worn plate of *Ostracion* from the Jackson (upper Eocene) of Clarke County, Alabama.

FAMILY DIODONTIDAE
porcupine fishes

Characteristically, these fishes have the body covered by sharp spines rising from a fluted base. The jaws bear a single pair of platelike teeth, often with successional teeth showing beneath and behind the functional tooth.

?*Diodon* sp. (fig. 51)

FIG. 51.   ?*Diodon* sp., mandible, BSC 784, Little Stave Creek, Clarke County, Alabama. x3.

White (1956) noted a specimen of this genus in the Enniskillen Collection but doubted that it came from Alabama. Jaws of this genus are not uncommon in the Gosport Sand at Little Stave Creek, Clarke County, Alabama. The teeth are bounded anteriorly by a parrotlike beak of bone. Tuomey (1850B, p. 162) noted this genus (as *"Diudon"*) from an unspecified Tertiary locality.

Class Amphibia

The amphibians are the most primitive of the land vertebrates. Their early representatives are barely distinguishable from the advanced crossopterygian fishes. The earliest record we have of amphibians is from the very Late Devonian of Greenland where two forms are found. *Elpistostege* is a poorly known type, only doubtfully amphibian, but from definitely Devonian rocks. *Ichthyostega* is a true amphibian with distinct marks of its fishy origins, including remnants of dorsal and caudal fins. Its remains come from slightly younger rocks, so close to the Devonian-Mississippian boundary that no one seems willing to say that the beds are of one age or the other.

The characteristic feature of the amphibians is their need to pass their early stages in water. The young amphibian (tadpole) is highly fishlike, with fins and gills. When it reaches a sufficient maturity, it rapidly transforms into a land animal—not a very efficient one in the light of later forms, but a land animal. A few amphibians have succeeded in bypassing the water-living stage by various subterfuges. The wormlike, burrowing caecilians have some African species reported to retain eggs within the body not only until after hatching but until metamorphosis. A few more normal amphibians essentially build little lakes in which to raise their young.

The living amphibians comprise the caecilians, salamanders, frogs, and toads. They are but poor remnants of the varied amphibian faunas of the Late Paleozoic. Then the class included, among others, powerful and aggressive carnivores up to 10 feet in length.

Tied as they are to the water for reproduction, it is not surprising that many amphibians spend most of their lives in or near water. A few have even given up the unequal struggle of being land animals in competition with their more advanced descendants and again have become totally aquatic. Some of these have even given up the need to metamorphose; these attain sexual maturity while retaining the characteristics of the larval stage. Such animals are termed *neotenic* and keep such features as external gills throughout life.

No fossil amphibian bones are yet known from Alabama. As a result, we know very little of the history of this class in the state.

We are not totally without information, however. In 1929, a remarkable series of amphibian trackways was collected by the Alabama Museum of Natural History and the Geological Survey of Alabama. These specimens were obtained at the Number 11 Mine of the Galloway Coal Company, near Carbon Hill, Walker County. The specimens came from a shale immediately overlying the Jagger Coal Seam of the middle part of the Pottsville Series (Early Pennsylvanian). These tracks were described and figured by T. H. Aldrich (*in* Aldrich and Jones, 1930).

The study of tracks is referred to as *ichnology*. This field cannot claim to be an exact science. Rather, it is an attempt to do the best we can with information that is incomplete and exceedingly difficult to interpret. This information can be important, if we can interpret it accurately. Tracks are characteristically found in places that yield little or no bone; thus they give us a look at information available nowhere else.

Aldrich (*in* Aldrich and Jones, 1930) attacked the problem of these tracks in a purely descriptive way. He described thirteen "species" of tracks, all new, based on variations in their shape, size, and spacing. These "species" were then arranged in eight "genera," seven of which were new. This procedure was valid in his day; it served the function of giving names to these tracks as a convenient means of further discussion.

Naming, however, is only the start of the work. Exactly what kind of animal made these tracks, and what can these tracks tell us about the habits of that animal? Unfortunately, it may not be possible to answer the first question, but some speculations seem to make sense. On the other hand, it is possible to learn considerably more about the habits of the animals and the conditions under which the tracks were formed.

All known four-legged amphibians (excluding the jumping forms) were short-legged animals that dragged the body and especially the tail while walking. Comparatively few of the Walker County trackways show any sign of a body or tail drag mark, though the tracks were made in soft mud. This absence of a drag mark probably indicates that the tracks were made in shallow water rather than on an exposed mudflat. The body and tail were floating; the feet were partly swimming yet pulled the animal along when they touched bottom. In such circumstances, slight differences in water depth could make considerable differences in just how much of the foot reached bottom. The "species" *Hydromeda fimbrata* and *Limnosaurus alabamaensis* of Aldrich are little more than claw marks made by animals that could barely reach bottom. They are thus quite uninformative about what type of animals made them. This lack of information eliminates two of the thirteen "species" from further consideration.

The specimen named *Ctenerpeton primum* by Aldrich is a set of markings made by a current of water around an object, probably a carcass, resting on the bottom. No remains of the object are present; if a carcass, it may well have bloated and drifted off. This is another "species" eliminated.

The specimens named *Trisaurus lachrymus* and *T. secundus* are better tracks than the first two considered above but are still uninformative. Only three toes and a faint pad impression are visible. Aldrich notes that the tracks of *T. secundus* "give the impression of the animal walking on its toes." This kind of track is precisely what one would expect in water slightly shallower than the depth in which simple claw marks are produced. The toes, perhaps not all, were completely on the bottom; part of the body of the foot was just touching.

This leaves eight "species" of tracks for discussion. We suspect that as few as three different species were involved and believe that we can make some speculations about more precise relationships.

?Order Rhachitomi

The Rhachitomi comprise an extremely prominent group of amphibians in the Late Paleozoic. Normally, they are distinguished by details of the vertebral and cranial structure. More important for our purposes is the structure of the feet. To note their distinctive feature, picture the tracks made by your hand if you placed it in soft mud. These tracks would show four digits (the fingers) roughly parallel to each other, and a fifth digit (thumb) at a sharply divergent angle. On tracks made by the two hands, the thumb marks would point toward each other.

This arrangement is not the case in the rhachitomes. There are five digits, one of which is divergent, as in your hand. But instead of being the first digit (thumb or big toe), the divergent digit is the fifth (little finger or toe). The "thumb" marks on a trackway of a rhachitome would then point out, not in. Two of the trackway types from Alabama show this pattern.

Attenosaurus subulensis Aldrich (fig. 52)

Attenosaurus subulensis Aldrich *in* Aldrich and Jones, 1930, p. 13, pls. 2–4.
Attenosaurus indistinctus Aldrich, *in* Aldrich and Jones, 1930, p. 13, pl. 1.

The trackways are large, up to 10 inches long and 7 inches wide. The trackway illustrated by Aldrich as Plate 2 (Plate 4 is a single track from this trackway) is designated here as the holotype of this track species. All known specimens are in The University of Alabama Museum of Natural History.

The digits are very long and slender with a divergent outer "thumb," indicating affinities with the Rhachitomi. Only the impressions of the hind feet are well preserved. Forefoot impressions consist of claw marks with only three toes shown. The digits were held nearly straight, except for the curved claw.

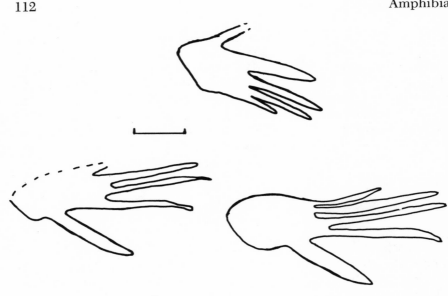

FIG. 52.   Amphibian tracks, Carbon Hill, Walker County, Alabama. *Attenosaurus subulensis* Aldrich. Three hind foot tracks. Redrawn from Aldrich and Jones, 1930. Scale bar 5 cm.

We are exercising a rarely used nomenclatural option here. The name *Attenosaurus indistinctus* has line priority over *A. subulensis* (*i.e.*, it occurs higher on the page where described originally; therefore it appeared in print first and has priority in use, technically). We are here exercising the option of the first revisers to ignore this priority, as *A. indistinctus* is a faint track whose preservation fits its name.

Cincosaurus cobbi Aldrich (fig. 53)

?Bipedes aspodon Aldrich *in* Aldrich and Jones, 1930, p. 23, pl. 5.
Cincosaurus Cobbi Aldrich *in* Aldrich and Jones, 1930, p. 27, pls. 6–7.
Cincosaurus Fisheri Aldrich *in* Aldrich and Jones, 1930, p. 27, pl. 8 (in part).
Cincosaurus jaggerensis Aldrich *in* Aldrich and Jones, 1930, p. 28, pl. 9.
Cincosaurus Jonesii Aldrich *in* Aldrich and Jones, 1930, p. 28, pls. 10–11.

All of these nominal "species" are based on tracks much smaller than those of *Attenosaurus*. The largest specimen (type of *Cincosaurus fisheri*) is 5 inches (13 centimeters) long; most specimens are about 2 inches long.

These specimens are very close in morphology, and the several "species" were distinguished largely on size and posture. As we have suggested, the posture could vary greatly with water depth. The type of *C. jonesii* consists of little more than claw marks. The other speci-

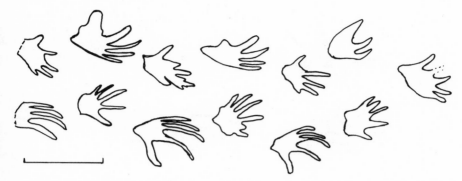

FIG. 53. *Cincosaurus cobbi* Aldrich. Trackway of fore and hind feet. Forefoot tracks are smaller and nearer midline. Redrawn from Aldrich and Jones, 1930. Scale bar 5 cm.

mens show five digits, all rather strongly curved, with the concave edge lateral on the first four and medial on the laterally placed "thumb." They probably represent a quite different animal from *Attenosaurus*, though the lateral "thumb" still indicates affinities to the Rhachitomi. The front feet are noticeably smaller than the hind feet.

  Bipedes aspodon may be a senior synonym of the species of *Cincosaurus*. The tracks referred under this name show only two toes, both strongly curved. They could well be tracks made by a swimming *Cincosaurus* in which only the leading edge of the foot touched bottom on each downstroke. There is no indication of a thumb; if the animal had a medial thumb as in normal amphibians, this should have touched bottom first. Therefore, it seems that *Bipedes* also was a rhachitome. As it is about the size of most *Cincosaurus*, it may well represent the same animal.

?Order Seymouriamorpha or ?Order Anthracosauria

  We are not able to distinguish tracks of these two orders. They are closely related, and the Seymouriamorpha seem to be descended from the Anthracosauria. Again, both orders are characterized mainly by vertebral structure.

  The Anthracosauria are very primitive amphibians confined to the Mississippian and Pennsylvanian. They tend to be long-bodied and short-limbed, indicative of largely aquatic habits.

  The Seymouriamorpha are named for the genus *Seymouria* from the Permian of Texas. This fascinating form is a "fence-sitter" with an almost equal mixture of amphibian and reptile characters. As a result, it and its relatives have been assigned to either class by various workers. Most modern opinion leaves the group in the Amphibia for various technical reasons.

Quadrupedia prima Aldrich (fig. 54)

Quadrupedia prima Aldrich *in* Aldrich and Jones, 1930, p. 53, pls. 8, 15.

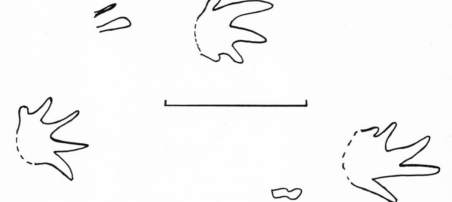

FIG. 54.   *Quadrupedia prima* Aldrich. Hind and very faint forefoot tracks. After Aldrich and Jones, 1930. Scale bar 5 cm.

Only hind-foot tracks are well enough preserved to show anything about this form. The forefeet are represented only by claw marks and impressions of two digits. Thus the animal had hind legs longer than front, not surprisingly, and was walking in water just deep enough to allow the hind feet full contact with the bottom and yet allow the forefeet only to scrape the mud.

The tracks are about 3 centimeters long and wide and show that the animal had a stride of about 13 centimeters. The toes are broad and show little trace of webbing. The first digit (big toe) is very stumpy. The other digits seem to have been rather mobile, as no two tracks on the type trackway show precisely the same relationships. There are also variations in track shape because of slippage.

Class Reptilia
reptiles

To most people, the first vision the word "fossil" brings to mind is of a gigantic scaly monster lumbering through a steaming swamp. Fossil remains of reptiles are among the most spectacular, and certainly the most popular, of vertebrate fossils.

How does one distinguish a reptile from any other vertebrate? In living forms, this is fairly easy. Only the amphibians could be confused with reptiles. The skin of a reptile is always scaly; that of an amphibian is slimy or warty. Contrary to a widely held notion, reptile skin is not slimy, except in the case of a water-living reptile with a thin coat of algae or moss.

Among fossil forms, the distinction is not so simple. The amphibians lead into reptiles with scarcely a break (see *Seymouria*, above). At the other end of this evolutionary scale, it is rapidly becoming impossible to draw a firm line between reptiles and mammals without using a purely arbitrary criterion.

The classification of the reptiles is based, at the subclass level, on the structure of the skull. In particular, it is based on the number and position of *temporal fenestrae* (fig. 55). These are openings in the skull roof that allow the main jaw muscles to bulge for greater efficiency. The amphibians and the primitive reptiles have a double skull, in a way. Internally, there is a fairly solid braincase with only the openings necessary for nerves, blood vessels, and the spinal cord. Externally is a completely separate roof formed in the underlayers (dermis) of the skin. The temporal muscles (the main jaw closers positioned between these two skull layers) must have room to expand to develop full efficiency. It is not surprising, therefore, that various groups of reptiles developed skull openings in several different ways for the bulging of the temporal musculature.

In the Subclass Anapsida, no true temporal openings are present, and the outer skull roof is more or less complete (see fig. 55A). This subclass comprises the primitive reptiles (Order Cotylosauria) and the turtles (Order Chelonia). Many of the turtles have found a functional replacement for temporal fenestrae by developing a notch in the posterior margin of the outer skull.

The Subclass Ichthyopterygia contains a single order, the Ichthyosauria. There is a single temporal opening very high on the side of the skull, with its ventral margin formed by the postfrontal and supratemporal bones. Some idea of the unusual nature of this arrangement may be gained from the name of one of these bones. The *supratemporal* was named because it normally lies above the temporal fenestra; here it is below. The ichthyosaurs are highly modified reptiles that are completely marine (see fig. 55B) and look much like fishes or porpoises. No ichthyosaurs are known from Alabama. Their heyday was during the Triassic, Jurassic, and Early Cretaceous, ages whose rocks are not known to be exposed in Alabama.

The Subclass Synaptosauria also has a single temporal opening high

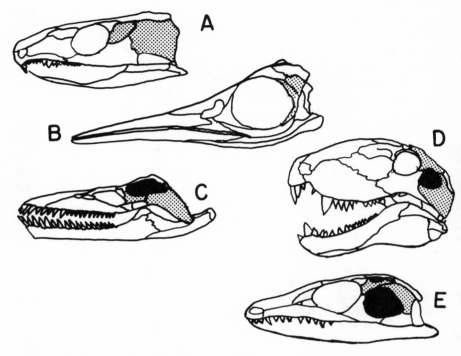

FIG. 55.   Reptilian skull types. Dark areas are temporal openings. Shaded areas are
postorbital and squamosal bones. Not to scale. After Colbert.

A: anapsid skull of *Captorhinus*, a cotylosaur. No temporal openings.
Found in primitive reptiles and in turtles. In the back of the skull, some
turtles develop a deep notch, never closed posteriorly by a bony bar, that
serves the purpose of a temporal opening.

B: parapsid skull of *Ichthyosaurus*. Temporal opening very high on skull,
barely visible in this view. The postorbital and squamosal form no part of its
margin. Found only in ichthyosaurs. Note huge orbit (eye socket). Sclerotic
plates (ring of stiffening bones in eyeball) omitted in this drawing.

C: euryapsid skull of *Muraenosaurus*, a plesiosaur. Single high temporal
opening whose *lower* border is formed by the postorbital and squamosal.
Found in plesiosaurs and their land and aquatic relatives.

D: synapsid skull of *Dimetrodon*, a pelycosaur. Single low temporal opening whose *upper* border is formed by the postorbital and squamosal. Found
only in mammallike reptiles and their mammalian descendants.

E: diapsid skull of *Youngina*, an eosuchian. Two temporal openings (upper rather small here) with postorbital and squamosal forming a bar *between*
them. Found in archosaurs (in which there would be an opening—the antor-
bital fenestra—in front of the eye) and in lepidosaurs (in most of which the
lower bar is lost, leaving the cheek open). *Youngina* is a primitive diapsid
that might belong to either subclass.

on the skull; however, its lower border is formed by the postorbital and squamosal bones in more normal fashion. There are two orders, the Protorosauria, a miscellaneous group of terrestrial reptiles with this skull pattern, and the Sauropterygia. This latter order includes marine reptiles such as the plesiosaurs, fragmentary remains of which are known from the Cretaceous rocks of Alabama (see fig. 55C).

The Subclass Synapsida also has a single temporal opening, but this is low down on the side of the skull, with the postorbital and squamosal meeting above it (see fig. 55D). This group grades almost imperceptibly into the mammals, which have a similar skull pattern. You can feel this in your own skull. Place your fingers just in front of the tops of your ears, and clamp your jaw shut. Feel the temporal muscles bulging? No synapsids are known in the fossil record of Alabama. They flourished from the Late Pennsylvanian through the Triassic, but no known rocks of these ages are exposed in Alabama.

The Subclass Archosauria has two temporal openings with the postorbital and squamosal bones meeting between them (see fig. 55E). Two other bones, the jugal and quadratojugal, form a bar beneath the lower opening. This subclass includes several orders: the Thecodontia (primitive archosaurs), Crocodilia (crocodiles and alligators, the only living archosaurs), Pterosauria (flying reptiles), Saurischia (some "dinosaurs," including all the carnivorous forms and some herbivores), and Ornithischia (the rest of the "dinosaurs," all of them herbivorous).

The Subclass Lepidosauria also has two temporal openings, like the archosaurs. It is difficult to distinguish early lepidosaurs from early archosaurs, but in the more advanced lepidosaurs the bony bar below the lower temporal opening is lost. This modification opens up the entire cheek region of the skull and allows the lower jaw joint to move fairly freely forward and back. This subclass contains the most familiar of the reptiles, the lizards and the snakes (combined in an Order Squamata), plus two obscure orders, the Eosuchia (primitive) and the Rhynchocephalia (still surviving in New Zealand). There is at present a growing tendency to place the lizards and snakes in separate orders. The terminology is variable. The lizards are placed in the Lacertilia or Sauria and the snakes in the Serpentes or Ophidia. The names are equivalent; the first of each pair is in Latin, and the second in Greek. We will follow a more conservative approach and use the Lacertilia and the Serpentes as suborders.

The bones of many fossil reptiles, particularly of marine forms, are abundant in the Cretaceous rocks of Alabama. A few land or freshwater varieties also are found there. Scattered remains of reptiles have been reported from the Tertiary rocks of Alabama.

Subclass Anapsida

These reptiles, as noted above, lack temporal fenestrae. The skull is either solidly roofed or the posterior margin is deeply notched to allow bulging of the jaw muscles. There are two orders. The Cotylosauria, a very primitive group probably ancestral to the rest of the reptiles, is unknown in the fossil record of Alabama. It is not totally inconceivable, however, that some of the tracks discussed (see above, Amphibia) belong to cotylosaurs. The other order, the Chelonia, has an abundant fossil record in Alabama and is unmistakable.

Order Chelonia
turtles

Turtles are common vertebrates ranging back in geologic time to the Triassic Period. Their habitats vary from vacant lots in New York City to piney woods in the South, from sluggish streams and lakes to deep marine environments. Most children have enjoyed baby turtles purchased from pet shops. It should be unnecessary to describe these unique animals, at least for their general features. However, the structure of the shell, as well as other details, will need some treatment.

The shell is composed primarily of bones normally present in the vertebrate body (fig. 56). It is divided into two parts: the *carapace,* or dorsal shell, and the *plastron,* the under shell on the belly. The carapace is composed of three basic series of bones. Along the center are eight *neural* plates fused to the underlying vertebrae. On each side of the neurals are eight elongate slats of bone forming the bulk of the carapace. These are fused to the underlying ribs (and perhaps are outgrowths of them) and are called either costals or pleurals. *Pleural* seems to be the preferable term, as the overlying horny scutes also are called costals. Around the margins of the carapace is a series of small bones, variable in number, appropriately termed *marginals.* Two additional terms are used if some of the marginals lie on the midline in line with the neurals. An anterior element of this type is called a *nuchal,* only one of which is ever present. Posterior elements on the midline in the marginal series are called *pygals,* of which there may be several.

Exterior to the bones of the carapace in most turtles lie the horny scutes mentioned above. Their margins do not coincide with those of the bones; rather, they overlap to give a sturdy interlocking structure. The scutes themselves are not preserved in fossil turtles, but their margins often are marked by visible grooves in the bones.

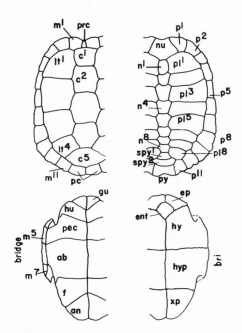

FIG. 56.   The turtle shell, based on *Pseudemys*. Left: scutes. Right: bony shell. Top: carapace. Bottom: plastron. No scale. Scutes: *ab*, abdominal; *an*, anal; c^{1-5}, centrals; *f*, femoral; *gu*, gular; *hu*, humeral; lt^{1-4}, laterals; m^{1-11}, marginals; *pc*, postcentrals; *pec*, pectoral; *prc*, precentral. Bony plates: *ent*, entoplastron; *ep*, epiplastron; *hy*, hyoplastron; *hyp*, hypoplastron; n^{1-8}, neurals; *nu*, nuchal; p^{1-11}, peripherals; pl^{1-8}, pleurals; *py*, pygal; spy^{1-2}, suprapygals; *xp*, xiphiplastron. After Carr, 1952.

The plastron usually is composed of elements that are much larger in size and fewer in number than those of the carapace. The plastron normally is much flatter than the carapace and consists of nine bones, eight of which lie in four pairs, each member of a pair joining its mate across the midline. From front to back, these bones are termed the *epiplastron*, the *hyoplastron*, the *hypoplastron*, and the *xiphiplastron*. The ninth bone, a diamond-shaped element called the *entoplastron*, lies in the region bounded by the epiplastra and the hyoplastra. The plastron usually is covered with scutes like those of the carapace. The region in which the plastron is joined to the carapace is appropriately called the *bridge*.

Only a few extremely primitive turtles have teeth, which are confined to the palate. The jaws are covered instead by a horny beak. The surface of the jaw normally bears small openings for blood vessels and nerves to the beak and usually is roughened for better beak attachment. The skull is generally quite massive, as befits the rest of the beast. The legs are usually stumpy, as are the feet in walking forms. The great sea turtles, however, may have the feet, especially the front, expanded into enormous paddles.

Many aquatic turtles have the shell considerably reduced and lightened. This modification is especially true of the sea turtles in which the shell may be reduced to a bare skeleton.

The classification of the turtles is based largely on the structure of the neck. The very primitive turtles (suborder Amphichelydia) do not seem to have been able to pull the head into the shell. The Suborder Pleurodira includes turtles that bend the neck into a sideways curve to withdraw the head; the Cryptodira use a vertical S-curve. No amphichelydian turtles are known from the fossil record in Alabama.

Suborder Pleurodira
side-necked turtles

As noted above, these turtles withdraw the head by a sideways curve of the neck. The earliest known fossils are from the Early Cretaceous of Europe; they survive today in Australia, South America, Africa, and Madagascar.

FAMILY PELOMEDUSIDAE

One of the two families of pleurodiran turtles that still survive, this group is more primitive than the other family, the Chelyidae, in that the plastron is composed of eleven bones; the extra two elements are along the midline behind the entoplastron and are called mesoplastra. The mesoplastra are also present in many amphichelydian turtles.

Bothremys barberi (Schmidt) (fig. 57)

Podocnemis barberi Schmidt, 1940, pp. 1–12, figs. 1–5.
Podocnemis alabamae Zangerl, 1948B, p. 25, pl. 4, figs. 3–13.
Bothremys barberi, Gaffney and Zangerl, 1968, pp. 198–208, figs. 2–12, 17, 18, 19D, 21E, 22D–E.

Other than this species, no turtle known from the Alabama Cretaceous possesses a complete shell, not reduced to a skeleton. Although the family is characteristically freshwater, specimens of this species have all come from marine rocks. There are numerous specimens, suggesting a nearshore marine habitat for the form.

The carapace is about two feet (60 centimeters) in length and breadth and slightly broader than long. Seen from above, it is nearly

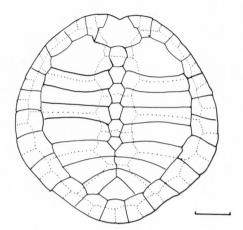

FIG. 57. *Bothremys barberi* (Schmidt). Carapace of type showing preserved scute (dotted) and plate (solid) margins. After Schmidt, 1940, fig. 2. Scale bar 10 cm.

circular in outline, but slightly notched anteriorly. The carapace shows some peculiar features. There are only six neurals, and the last two costals meet at the midline. There are eleven marginal plates, plus a nuchal and a pygal. The plastron shows a deep notch posteriorly.

Bothremys barberi is known from Upper Cretaceous rocks in New Jersey, Alabama, Arkansas (type specimen), and Kansas. Gaffney and Zangerl (1968) recognize three subspecies, A from the Atlantic Coast, B from the Gulf Coast, and C from Kansas. The Alabama form should be referred to as *B. barberi barberi* (Schmidt), as it is almost identical to the type specimen from Arkansas.

Suborder Cryptodira

Most living and fossil turtles belong to this suborder. The head is withdrawn by placing a vertical S-curve in the neck. A number of forms, especially the sea turtles, have lost the ability to completely withdraw the head.

We must take time here to recognize gratefully the work of Dr. Rainer Zangerl of the Field Museum of Natural History in Chicago. For many years he has been the principal driving force behind investigations of the Cretaceous vertebrates of Alabama. His contributions to the knowledge of fossil turtles have been particularly significant. Much of our treatment here is based on his work.

FAMILY DERMATEMYDIDAE

Dermatemys mawi, the only surviving species of this family today, is found on the east coasts of Mexico and Guatemala. It was abundant and varied in the Late Cretaceous and Tertiary of North America.

Agomphus alabamensis Gilmore (fig. 58)

Agomphus alabamensis Gilmore, 1919C, pp. 123–125, pl. 35, text-fig. 5.

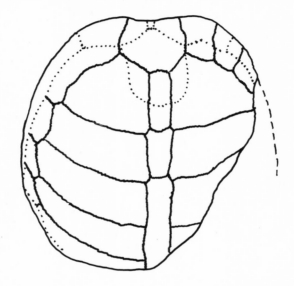

FIG. 58.   *Agomphus alabamensis* Gilmore. Carapace showing preserved scute (dotted) and plate (solid) margins. Type, USNM 8806, Moscow Landing, Sumter County, Alabama. After Gilmore, 1919C. x¼.

This turtle is one of only three vertebrate species yet known from the Paleocene of Alabama (the other two are sharks). It has a much more vaulted shell than that of any other known Alabama fossil turtle. The carapace, besides being unusually high, is rather narrow in top view. A peculiarity of the family is the shortening of the neural series so that some of the posterior costals meet along the midline. However, there are always eight neurals rather than seven as in the Pelomedusidae. The shell of this species is a little over a foot long.

The type specimen was collected on the Tombigbee River at Moscow Landing, Sumter County, Alabama, by E. H. Sellards in 1908. The locality is probably just north of the mouth of Sucarnoochee Creek and would be low in the Midway Group. Only about the front half of the shell is known.

FAMILY TESTUDINIDAE
tortoises

These represent the most thoroughly terrestrial family of the turtles. The shell is highly vaulted, though not so high as that of the dermatemydids. The neural series is normal, and the costals never meet along the midline. Some fossil forms attain very large sizes. *Colossochelys* from the Pliocene of India attains carapace lengths of nearly 8 feet (over 2 meters). Surviving giant tortoises are confined to the Galapagos and to some islands in the Indian Ocean; however, they were common in North America during the Pleistocene. The oldest tortoises are found in Eocene rocks.

Hadrianus? schucherti Hay (fig. 59)

Hadrianus schucherti Hay, 1899D, p. 22, pls. 4–5.
Hadrianus? schucherti, Hay, 1908A, pp. 382–383, fig. 481.

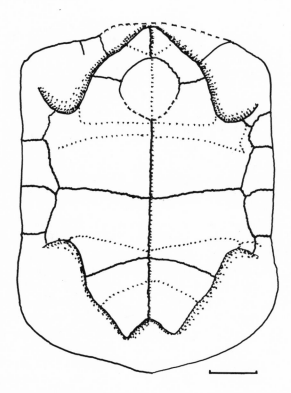

FIG. 59.   *Hadrianus? schucherti* Hay. Plastron of type, USNM specimen. After Hay, 1908. Scale bar 10 cm.

This species is based on a specimen collected by Charles Schuchert in the Jackson Group of Choctaw County, Alabama. The preserved parts consist of a nearly complete plastron and most of the marginal series. The top of the carapace was eroded away. Because *Hadrianus* is distinguished from *Testudo* by the shapes of the neural plates, this specimen cannot be placed definitely in either genus.

The shell is over 2 feet long (675 millimeters). In top or bottom view, it is nearly rectangular. The sides are parallel and almost straight; both ends are convex. The posterior end of the plastron is deeply notched, and the entoplastron is almost circular.

The specimen was a male, as indicated by the concavity of the center of the plastron (Hay, 1908A). This is the only turtle known from rocks of late Eocene age in Alabama. It is one of many specimens collected near the Old Cocoa Post Office in Choctaw County, probably the most famous vertebrate fossil locality in Alabama.

FAMILY PROTOSTEGIDAE

An early family of marine turtles, this group is known only from Upper Cretaceous rocks. Members are closely related to the Toxochelyidae, another Late Cretaceous family, from which they differ in several features. The carapace is much reduced in both families by failure of the pleural plates to unite near the margin. In the Protostegidae, there is a complete marginal ring on which the end of the ribs insert in notches or shallow pits, never in deep sockets as in the Toxochelyidae. The plastron also is much reduced by failure of its bones to meet. The main plastral bones (hyoplastra and hypoplastra) are massive plates of bone with long projections around their edges. These projections formed a nearly complete basket under the body but left the plastron with large openings, called *fontanelles,* on all sides of each bone. The entoplastron is a large T-shaped bone, but the epiplastra are reduced or absent. The xiphiplastra are small but meet at the midline, enclosing a large central fontanelle in the midst of the plastron in a complete ring of bone. The eyes are placed rather far back in the skull for a marine turtle, most of which have the eyes placed far forward.

Protostega dixie Zangerl (fig. 60)
Protostega dixie Zangerl, 1953A, pp. 94–118, pl. 7, text-figs. 30–55.

In this large species, with a carapace length of about 5 feet (1.5 meters), the carapace structure is much reduced, with bare remnants of the expansion of the pleural plate around the ribs. There are nine

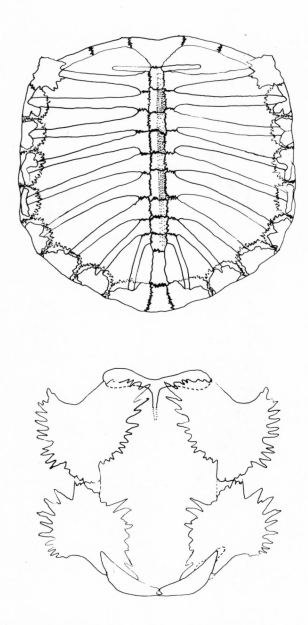

FIG. 60.   *Protostega dixie* Zangerl. Reconstructions of carapace and plastron, based on various specimens from Dallas, Greene, and Hale counties, Alabama. Compare with *Bothremys barberi* (fig. 57). Note here that sutures between carapace bones are very elaborate, giving the greater mechanical strength required by this very skeletonized shell. Dotted outlines show expansions of the marginals on the ventral surface. After Zangerl, 1953. *c.* x1/20.

neurals, the second, fourth, and sixth of which bear a sharp median crest. The third and fifth are saddle-shaped.

A common and widespread species in the Mooreville Chalk (Upper Cretaceous), specimens have been collected in Dallas, Hale, and Greene counties, Alabama.

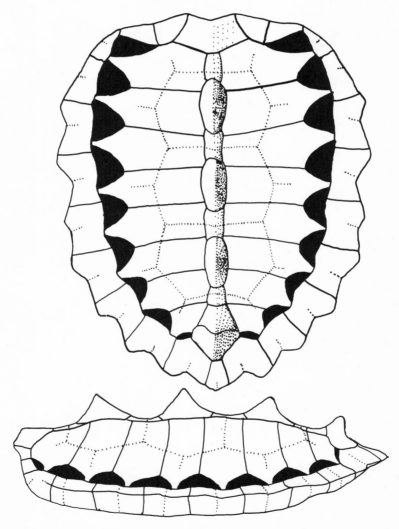

FIG. 61.   *Calcarichelys gemma* Zangerl. Reconstruction of carapace, dorsal and lateral views. After Zangerl, 1953A. x½.

Calcarichelys gemma Zangerl (fig. 61)

Calcarichelys gemma Zangerl, 1953A, pp. 119–123, figs. 56–58.

The carapace in this small species is less than a foot long. It is quite distinctive and unmistakable. The carapace is much less reduced than that of Protostega, with the pleural plates fused for about three-fourths of their length. The eight neurals are of unique form. The second, fourth, and sixth each bears a tall point of bone placed about at the midpoint of the plate. There is a fourth point posteriorly, partly on the eighth neural and partly on one of the pygals. The odd-numbered neurals are nearly flat. The marginals are strongly serrated, giving a sawlike edge to the carapace.

Only two specimens are known. The type is from near Burkville, Montgomery County, and the other from near West Greene, Greene County, Alabama. Both are from the Mooreville Chalk.

Chelosphargis sp. cf. C. advena (Hay) (fig. 62)

?Protostega advena Hay, 1908A, p. 199, text-figs. 256–259.
?Chelosphargis advena, Zangerl, 1953A, pp. 80–88, figs. 21–26.
Chelosphargis cf. advena, Zangerl, 1953A, pp. 118–119.

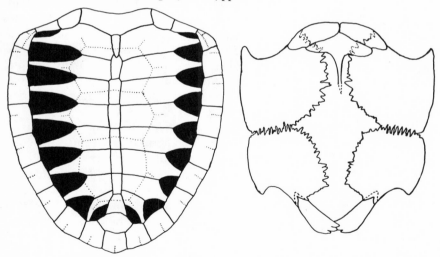

FIG. 62. Chelosphargis advena (Hay). Reconstruction of carapace and plastron. On carapace, left side shows juvenile with relatively larger fontanelles, right side adult. Not to scale, but known specimens range in carapace length from about 20 to 30 cm. After Zangerl, 1953A. Alabama material known to date can be referred to this species only doubtfully.

The occurrence of this species is doubtful in Alabama. Zangerl (1953A) notes that three specimens from the Mooreville Chalk of

Greene and Dallas counties are poorly preserved but show affinities
to Kansas specimens of this form. He also notes considerable differences, which perhaps indicate the presence of a new species as yet too
poorly known to describe and name. Our description and figure are
based on Kansas materials.

The carapace ranges, in different specimens, from 1 to 3 feet in
length (30 to 90 centimeters). It is somewhat heart-shaped in outline.
The neurals are not distinctly keeled and are of the same height,
forming a low, even ridge along the back. Most of the eight neurals are
nearly rectangular. The edge of the marginal series is nearly smooth.

FAMILY TOXOCHELYIDAE

Another family of Late Cretaceous marine turtles, the Toxochelyidae can be distinguished from the Protostegidae most readily by
the structure of the plastron. Instead of being composed of four almost
separate elements making contact only by long, almost spinelike processes as in the Protostegidae, the Toxochelyidae have a much more
solid structure that is roughly cross-shaped. The entoplastron is a
small, wedge-shaped bone. The hyoplastra and hypoplastra are su-
tured broadly together and extend far to the side, forming a very wide
bridge. The epiplastra are small and slender, while the xiphiplastra
are much larger; both are joined along the midline; thus completely
enclosing a large central fontanelle. This fontanelle is narrow with a
very ragged margin between the main plastral elements but expands
in the center into a large, smooth-edged opening. The plastron is
much like that of the living snapping turtles (Chelydridae).

The carapace is much like that of the Protostegidae. There are three
subfamilies: the Toxochelyinae, with no median keel and a smooth
edge on the carapace; the Osteopyginae, unkeeled like the previous
subfamily, with the peripheral margin smooth or weakly serrate, but
with the fontanelles of the carapace small or absent; and the
Lophochelyinae, with a powerful, serrated, median keel and a serrated periphery. The first and third of these subfamilies are known
from the Upper Cretaceous Mooreville Chalk of Alabama.

Subfamily Toxochelyinae

See above for the distinctions of this subfamily.

Toxochelys moorevillensis Zangerl (fig. 63)

Toxochelys moorevillensis Zangerl, 1953B, pp. 186–193, figs. 74–76, 70 (in part), pls. 14,
22–23.

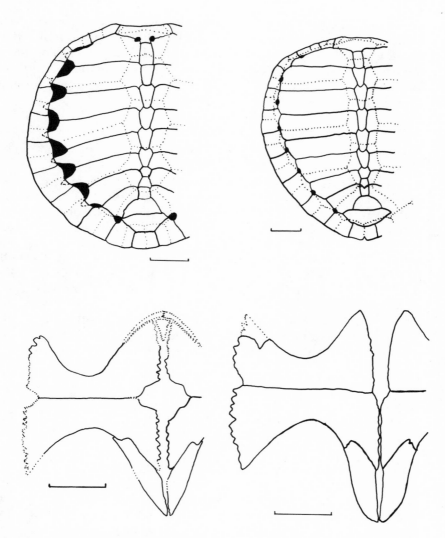

FIG. 63.   Two Alabama toxochelyid turtles. Left: *Toxochelys moorevillensis* Zangerl.
Right: *Thinochelys lapisossea* Zangerl. After Zangerl. Scale bars 10 cm.

This species is based on about sixty specimens collected over much
of the western part of the Black Belt of Alabama. Zangerl (1953B) lists
material from Dallas, Greene, Hale, and Lowndes counties. A typical
carapace is about 2 feet (60 centimeters) long, and about as wide as
long. The outline is somewhat cordiform (heart-shaped) with a shallow anterior notch and a slightly pointed posterior end. There are nine
neurals, and each is almost flat, mostly in the shape of a blunt ar-

rowhead pointing posteriorly, with a notched base. In most specimens, the neurals are about equal in length and width, rather than
being wider than long as in other species of *Toxochelys*. The fontanelles are rather small. There are small anterior fontanelles between
the nuchal and first pleural, with the first neural sometimes forming
part of their border.

Thinochelys lapisossea Zangerl (fig. 63)

Thinochelys lapisossea Zangerl, 1953B, pp. 200–202, fig. 82, pl. 25.

Only four specimens of this species are known, all from near Harrell
Station, Dallas County, Alabama. The type is about 2½ feet (70 centimeters) in carapace length, but there are indications of larger individuals. As in other toxochelyines, the neurals are unkeeled. They are
much longer than wide, unlike *Toxochelys*. The carapace is rounded
anteriorly, widest at the second and third pleural, and distinctly
pointed posteriorly. The fontanelles are very small openings tucked
between the end of a pleural and the junction of two marginals. There
is no trace of the small anterior fontanelles between the nuchal and
the first pleural that occur in *Toxochelys*. The shell is unusually
heavy, and Zangerl (1953B) notes that all known specimens, with the
exception of a small individual, have "a bluish, opaline appearance."

Subfamily Lophochelyinae

See above for the distinctions of this subfamily.

Lophochelys venatrix Zangerl (fig. 64).

Lophochelys venatrix Zangerl, 1953B, pp. 224–226, figs. 95–97.

This is the first of the toxochelyids with keeled neurals that we shall
consider. The carapace is about 1½ feet (45 centimeters) long and
slightly wider (50 centimeters) than long. The edges are serrated, with
a single point on each of the marginals. There is a deep notch in the
front of the carapace, formed by a deeply curved nuchal. There are
nine neurals, most of which are evenly hexagonal, and each of which
bears a distinct keel. However, there are no extra bones (epineurals)
resting above and between the neurals forming spikes on the keel, as
in the other Mooreville lophochelyines. The fontanelles are moderately large.

All four known specimens are fragmentary. Three are from near
Harrell Station, Dallas County, and one from Mt. Hebron, Greene
County, Alabama.

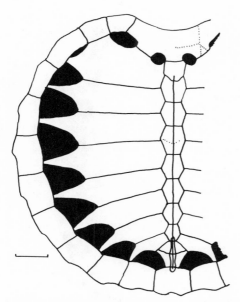

FIG. 64. *Lophochelys venatrix* Zangerl. Reconstruction of carapace. The peripheral bones have been crushed and were not originally as wide as shown. After Zangerl, 1953B. Scale bar 5 cm.

Ctenochelys tenuitesta Zangerl (fig. 65)

Ctenochelys tenuitesta Zangerl, 1953B, pp. 230–237, figs. 100–106, 122.

There are two species of *Ctenochelys* described from the Mooreville Chalk of Alabama. We will discuss the features of the genus first, then the characters that separate *C. tenuitesta* from *C. acris*.

The carapace of *Ctenochelys* is long and narrow compared with that of the other lophochelyines. The margins are strongly serrate, and the nuchal is more or less deeply curved. The lateral fontanelles are moderately large, and the ribs that extend beyond the ends of the pleural plates are slender. There is a pair of anterior fontanelles, bounded by the nuchal, the first pleural, and the first neural. The neurals are hexagonal and strongly keeled. Between neurals 2–3, 4–5, and 6–7 are prominent humps formed from extra bones (epineurals). The plastron is poorly known in both Alabama species.

In *C. tenuitesta* the keel humps are fairly low, though sharp. There is a fourth hump far posteriorly, resting on a suprapygal, immediately anterior to the pygal bone. All the rib processes on the pleural plates, except the first, are directed posteriorly. The outline of the carapace shows a less curved nuchal and a more pointed posterior end than in

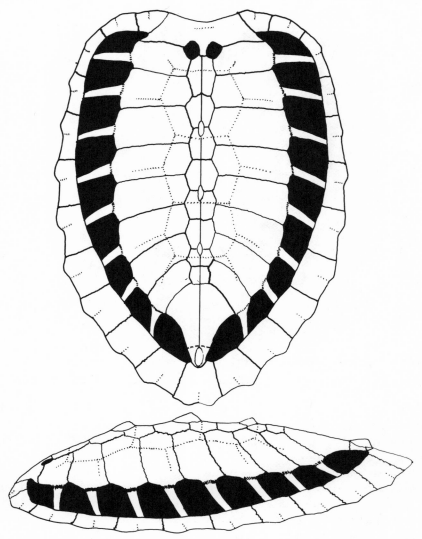

FIG. 65. *Ctenochelys tenuitesta* Zangerl. Dorsal and lateral views of restored
carapace. Note that carapace knobs are formed by separate bones (epineurals) rather than by the neurals as in the superficially similar protostegid
Calcarichelys. Total length of carapace of an adult, about 50 cm.

C. acris. A typical specimen is 1½ feet (45–50 centimeters) in carapace
length.

This species is one of the more common turtles in the Mooreville
Chalk. Specimens are known from Greene, Dallas, and Hale counties,
Alabama.

Ctenochelys acris Zangerl (fig. 66)

Ctenochelys acris Zangerl, 1953B, pp. 242–247, figs. 112–113.

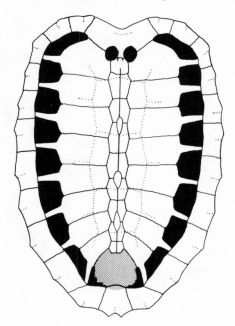

FIG. 66.   *Ctenochelys acris* Zangerl. Reconstruction of carapace, length of adult
carapace *c.* 65 cm. Structure of stippled area in suprapygal region unknown.
After Zangerl, 1953B.

Although found at the same localities, *Ctenochelys acris* is much
rarer than *C. tenuitesta*. Fewer specimens are known, and most are
fragmentary, so our knowledge of this species is far from satisfactory.

An extra element in the neural series, termed the *preneural*, is located between the nuchal and the first neural and replaces the first
neural in the border of the anterior fontanelles. The epineural humps
are taller and sharper than in *C. tenuitesta*. The nuchal is much more
deeply curved, and the posterior end is less pointed. An easily recognizable feature is the straightness of the rib processes of the pleural
plates. These are mostly directed straight laterally rather than posteriorly as in *C. tenuitesta*. The suprapygals are not known and may
have been much reduced. The type specimen is much larger than that
of *C. tenuitesta*, nearly 2 feet (60 centimeters) long.

Prionochelys matutina Zangerl (fig. 67)

Prionochelys matutina Zangerl, 1953B, pp. 254–258, figs. 114 (part), 118–120.

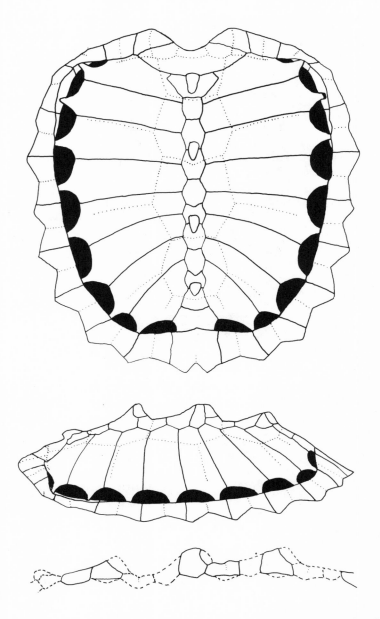

FIG. 67.   *Prionochelys*. Top and center: restoration of the carapace, dorsal and lateral views, of *P. nauta* from Arkansas. Bottom: lateral view, keel of *P. matutina* from Alabama, based on type, FMNH P 27561. After Zangerl, 1953B. Carapace length of adult of either species, somewhat over 1 meter.

Because this species is poorly known, our description will be based on the better preserved material of *Prionochelys nauta* from Arkansas, with notes on differences between *P. matutina* and *P. nauta*.

The shell ranges up to about 20 inches (50 centimeters) in length and is much broader, proportionately, than that of *Ctenochelys*. *Prionochelys* resembles *Lophochelys* in this respect but has a very prominently humped keel with large epineurals. The lateral fontanelles are comparatively small, and mature specimens have no anterior fontanelles. There are a preneural and nine neurals. The keel bears four or five prominent humps, the first on the preneural, then between neurals 2–3, 5–6, and 8–9. The fifth hump, if present, is between the pygal and the suprapygal. The neurals are considerably involved in the formation of the humps, unlike *Ctenochelys* in which the humps are almost completely formed by the epineurals.

There is no fifth hump in *P. matutina*, as far as is known; the top of the third hump is formed by the fifth neural, while the epineural in this location is confined to the posterior slope of the hump. The front of the shell is deeply indented, the inward curve involving the first marginals as well as the nuchal.

FAMILY DERMOCHELIDAE
leatherback turtles

This is the strangest living family of sea turtles, if not of all the turtles. There is only one living species, *Dermochelys coriacea*, found in all tropical seas and occasionally straying into colder waters as far north as Maine and England. It is the largest of the living turtles, attaining a length of at least 7½ feet (2.2 meters) and a weight of 1200 pounds (about 500 kilograms).

The shell is reduced to scattered, irregular plates of bone in the skin. There are no horny scutes; rather the skin of the back is tough and leathery (hence the common name). The oldest known fossil representatives are Eocene in age. The family is so different from the other turtles that some have considered it a totally independent development, unrelated to, but paralleling in many features, the true turtles. Others have considered this the most primitive living turtle. More commonly held at present is the idea that the leatherbacks are simply sea turtles that have carried the loss of armor to its extreme (Romer, 1945).

Fragments of a probable dermochelid were noted from the Eocene of Alabama by Müller (1849A, p. 34, pl. 27, fig. 7), thus being the first fossil turtle remains found in this state. Winge and Miller (1921A,

p. 56) referred this material questionably to the fossil genus *Psephophorus*. The fragments are rather nondescript and uninformative and will not be discussed further here.

FAMILY CHELONIIDAE
sea turtles

The living forms of this family are abundant and varied, though their numbers are now severely reduced by hunting, egg collecting, and destruction of nesting grounds. They include the green turtle of turtle-soup fame *(Chelonia);* the hawksbill, whose shell scutes yield the prized tortoiseshell *(Eretmochelys);* the loggerhead, whose large head and powerful jaws can sometimes bite off boat oars *(Caretta);* and the small and little-known ridleys *(Lepidochelys)*. People have long enjoyed the food gained from the eggs and flesh of these turtles and the ornaments and jewelry from their shells. We are now in danger of wanting too much of a good thing. For more information on the habits of these remarkable animals, see Carr (1955, 1973).

Fossil material of cheloniid turtles dates as far back as the Early Cretaceous. Although Tertiary specimens are known from the Atlantic Coastal Plain, none seem to have been collected in Alabama. There is, however, a remarkable specimen from the Mooreville Chalk of Alabama.

Corsochelys haliniches Zangerl (fig. 68)

Corsochelys haliniches Zangerl, 1960, p. 286ff., pls. 30–33, figs. 125–145.

One of the larger turtles of the Mooreville Chalk, the type specimen has a carapace length of about 4 feet (1.15 meters). The outline of the carapace immediately distinguishes this species from any of the other known Mooreville turtles. In the other forms, the front of the carapace is straight or indented above the neck. *Corsochelys* has a prominent flange of bone, mainly formed from the nuchal, over the neck, and a similar flange posteriorly, over the tail. There seems to be connection between the pygal (over the tail) and the suprapygals (at the posterior end of the neural series). There are eight neurals. Each corner of the carapace, over the limbs, is deeply excavated to give the flippers more freedom of movement.

The plastron is much reduced and generally reminiscent of that of the protostegids in that all the main elements bear long fingers of bone projecting on almost all sides.

The specimen was collected near West Greene, Greene County, Alabama.

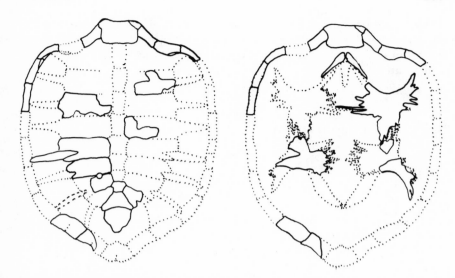

FIG. 68.   *Corsochelys haliniches* Zangerl. Reconstruction of carapace and plastron.
Only those parts shown in solid outline are preserved in the single specimen
(FMNH PR 249) known. Such an ambitious reconstruction can only be attempted on the basis of careful study of all available information, including
the way various parts work in related forms. Since this work was done,
another specimen, even more fragmentary, has been given to Auburn University. The new data do not contradict this restoration. Length of adult
carapace about 1.15 m. After Zangerl, 1960. The round hole between left
costals 6 and 7 is due to a parasite infection.

FAMILY TRIONYCHIDAE
soft-shelled turtles

Almost every fisherman in the South is familiar with this family.
These turtles are often called "soft-shell snappers," a term that describes them accurately but leads to confusion with the true snappers
(Chelydridae). Another description is more concerned with habits
than morphology: "turtles with long necks and short tempers." The
average fisherman, well acquainted with their ability at bait-stealing,
is more likely to describe them as "&?!$/&!/ turkles."

The anterior and posterior margins of the shell lack a continuous
series of marginal bones and are simply leathery flaps. The pleural
bones are shaped like those of many sea turtles: flat plates with a rib
process pointing laterally. The surface of these plates is highly ornamented with rows of shallow pits. The present tendency is to include the North American forms in the genus *Trionyx*, but much of the
literature uses the name *Amyda*.

Whitmore (*in* Isphording and Lamb, 1971) notes the presence of an unidentified *Trionyx* in the Pliocene Citronelle Formation in northern Mobile County, Alabama. Subfossil specimens are not uncommon on Indian sites.

Subclass Synaptosauria

Order Sauropterygia

Suborder Plesiosauria
plesiosaurs

A varied group of moderately large to gigantic marine reptiles, the plesiosaurs are found in rocks ranging in age from Triassic to Late Cretaceous. The body is flattened and turtlelike, without the solid armor of the turtles. The stubby tail apparently was not used for swimming to any great extent. The main motive power was supplied by four large flippers or paddles. These paddles are of remarkable shape, considering some modern-day sporting developments.

There was a recent revolutionary development in rowing races. A German rowing club adopted a new type of oar; instead of being blunt, the end of the oar was long and pointed. Their crews were soon beating everybody in sight until the new "leaf-shaped" oars became more widespread. Overlooked in the admiration for the scientific hydrodynamic studies by the Germans were two curious facts. When the oars of primitive tribes are examined, they immediately appear to have the same shape. Primitive man knew more of the practical hydrodynamics of rowing than we did until very recently; he adopted the shape we now know to be the most efficient. However, primitive man made this discovery as a latecomer in comparison to the plesiosaurs! The paddles of the most primitive plesiosaurs known have exactly the same shape as the scientifically designed German oars. Perhaps we should pay more attention to the lessons we can learn from nature. The hard school of nature taught this particular lesson to the plesiosaurs some two hundred million years before we finally caught on.

The most likely finds of plesiosaurian remains are isolated vertebral centra. Their size and presence in Cretaceous rocks should immediately distinguish them from anything except the mosasaurs (see below). They can be readily separated from mosasaur vertebrae by a quick look at the ends. If both ends are slightly dished, they are plesiosaurian. If one end is dished and the other a rounded ball, they are mosasaurian. If both ends are deeply dished, in particular if there is a hole through the center, think again. It is possible that you have the vertebra of an ichthyosaur, none of which has been found in Alabama. However, it is most likely a large fish, either shark or *Xiphactinus* (*q.v.*).

The teeth of the plesiosaurs are nearly round in cross section and rather slender, though usually with a blunted point. The enamel surface may be either smooth or weakly fluted (if strongly fluted, it may belong to *Polyptychodon,* a genus known in Texas but not in Alabama). There is never a sharp cutting edge.

Paddle bones are difficult to distinguish from those of mosasaurs. They are spool-shaped, with slightly expanded ends. The ends generally are less expanded relative to the shaft than those of the mosasaurs. If the bones are preserved together in their original position, the fingers are closely spaced and nearly parallel in plesiosaurs, but widely spread in most mosasaurs (there is a single exception in the mosasaur *Kolposaurus,* known only from California).

The only plesiosaur remains yet noted from Alabama are extremely fragmentary.

SUPERFAMILY PLESIOSAUROIDEA
plesiosaurs (in a strict sense)

These are plesiosaurs with long necks and small skulls. The neck, in advanced forms, may make up nearly half the total length. *Elasmosaurus* is the most extreme type. Out of a total length of about 40 feet, 18 feet is neck. The number of neck (cervical) vertebrae may be as high as 75. In this form, the skull is less than 2 feet long.

The reptiles of this superfamily seem to have been largely surface swimmers, lolling about on the surface and darting their heads after comparatively small fish (up to about 2 feet). Welles (1943) has closely examined the neck vertebrae of members of the group and suggests that some could bend the neck sideways through nearly two full circles. The largest of the plesiosaurs has a length of about 44 feet.

FAMILY ELASMOSAURIDAE
Cretaceous plesiosaurs

The character by which this family is distinguished from the Jurassic Plesiosauridae is the attachment of the ribs in the neck to the neck vertebrae. The Jurassic types have two heads on each cervical rib, and thus two pits on each side of the cervical vertebrae for attachment. The Cretaceous Elasmosauridae have only single-headed cervical ribs. In this family, the vertebrae are normally about as long as they are high and wide. The only exceptions are in the tail region of some forms.

Discosaurus vetustus Leidy

Discosaurus vetustus Leidy, 1851G, p. 326.
Discosaurus vetustus, Leidy, 1865A, pl. 5, figs. 4–6.
Cimoliasaurus vetustus, Cope, 1869A, p. 266.
Discosaurus vetustus, Hay, 1902A, p. 454 (type restricted to Alabama specimen); Welles, 1952, pp. 110–111 (indeterminate); Welles, 1962, p. 39 (indeterminate).

The literature on this species at this point probably outweighs the specimen. The type consists of two battered centra. One, at least, was collected by M. Tuomey and presented to the Academy of Natural Sciences of Philadelphia before 1856. It bears the label "Choctaw Bluff, Clarke County, Alabama," which is certainly an error: this locality is a well-known fossil site but is of Eocene age, millions of years after the extinction of the plesiosaurs. Another Choctaw Bluff, in Greene County, is rich in fossil vertebrates and is of Cretaceous age (upper Eutaw and basal Mooreville). It is probably the locality from which this specimen came.

The vertebrae are much wider and higher than long. This characteristic normally would suggest referral to the family Polycotylidae (see below). Welles (1952, 1962) followed this course but considered the specimen indeterminate (*i.e.*, not identifiable to species). We wholeheartedly agree with the latter conclusion. However, Welles (letter to Thurmond, 1971), while still considering the material indeterminate, considers it to be two anterior caudal centra of an elasmosaurid near *Alzadasaurus* (see Welles, 1952). He reports that the centra of the two forms are almost identical.

The authors know of two other occurrences of normal elasmosaurid material in Alabama, neither of which has been reported in the literature. One, in the collection of the Geological Survey of Alabama, is a single centrum found in Sumter County 15 feet deep in clay on the Tombigbee River at the crossing of the Alabama, Tennessee, and Northern Railroad by persons digging a bridge footing. It is indeterminate, but is probably elasmosaurian. A private collector has most of an elasmosaur neck, collected near Sawyerville, Hale County, Alabama.

SUPERFAMILY PLIOSAUROIDEA
pliosaurs

These reptiles are the short-necked plesiosaurs. The skull is much longer than that of the true plesiosaurs, and the neck is much shorter, usually only a little longer than the skull. Pliosaurs seem to have been more active hunters than the plesiosaurs, chasing down and killing larger prey, both fishes and smaller reptiles.

Some members of the pliosaurs attained huge sizes. A complete skeleton of *Kronosaurus* from the Lower Cretaceous of Australia is about 55 feet long, with a 10½-foot skull. There are indications of even larger specimens, with 12-foot skulls and teeth with crowns a foot long. Some British Jurassic specimens are scarcely smaller. One recent find (A. J. Charig, oral communication to Thurmond, 1967) spanned 18 feet across the hind paddles.

FAMILY POLYCOTYLIDAE

The polycotylids are Cretaceous pliosaurs, especially common in Late Cretaceous rocks. Strangely enough, the character that separates them from the Jurassic and Early Cretaceous Pliosauridae is the same one used to distinguish the Plesiosauridae from the Elasmosauridae: the cervical ribs are single-headed rather than having two heads.

To date (1980), no polycotylids have been described from Alabama rocks, though *Discosaurus* (see above) was long included in this family. Welles (letter to Thurmond, 1971) presently is working on a skeleton of this family from Alabama.

Another probable polycotylid from Alabama, unless this specimen belongs to *Discosaurus*, is a single centrum from Jasmine Hill, southeast of Wetumpka, Elmore County. The specimen was collected from a block of Mooreville Chalk that had been driven down into the underlying Tuscaloosa Group by what seems to have been a meteorite impact (Bentley, et al., 1970). Considering what the specimen has been through, it is not surprising that it is poorly preserved.

Welles (1962) calls this family Dolichorhynchopidae, an opinion challenged by Thurmond (1968) at Welles's suggestion. Welles (letter to Thurmond, 1971) now uses the older name, Polycotylidae.

Vertebrae of polycotylids normally can be distinguished from those of elasmosaurids by the fact that they are considerably shorter than broad or high.

Subclass Archosauria
"ruling reptiles"

See discussion of Reptilia, above, for the distinguishing characteristics of this subclass.

Order Crocodilia
crocodiles, alligators, gavials

These familiar animals are the only living archosaurs. Their fossil record goes back into the Triassic. Three living families are usually recognized, although some workers use only a single family. The Crocodilidae include the living crocodiles and their fossil relatives. The snout is relatively long and narrow, and the fourth tooth of the lower jaw fits into a notch in the side of the upper jaw. There are living crocodiles in all continents except Europe and Antarctica, though they reach North America only in southern Florida.

The Alligatoridae are more familiar to Alabamians, as specimens are found increasingly in the southern parts of the state. The snout is short and broad, and the fourth tooth of the lower jaw fits into a deep pit in the upper jaw. Alligators and their close relatives, the caimans, are found only in the Americas and, oddly enough, in the delta of the Yangtze River, China.

The Gavialidae have an extremely narrow snout and are entirely fish eaters (the others seem to eat anything meaty that comes along). They are presently confined to India and Burma.

Only three occurrences of fossil crocodilians are known from Alabama. Whitmore (*in* Isphording and Lamb, 1971) reports fragmentary specimens from the Pliocene rocks in northern Mobile County. The authors also have recovered scraps from the basal Gosport Sand at Little Stave Creek, Clarke County. These are of middle Eocene age. In these two cases it has been impossible to say anything more than just "crocodilian." The third specimen, discovered by one of the authors in 1973 in a local basement, was collected by a group of Boy Scouts in the early 1950s near West Greene, Greene County. These crocodilian remains are the best known from the Cretaceous of Alabama. The specimen consists of the complete pelvic girdle, entire right rear leg and foot, part of left rear leg, approximately thirty vertebrae (mostly caudal), fifty to sixty dermal scutes, and some rib fragments. Samuel W. Shannon, formerly of the Geological Survey of Alabama, studied this specimen (AGS 65A V-1096) and shared casts and photographs with Langston at the University of Texas. Langston (personal communication to Shannon) assigns this form to *Deinosaurus* (*Probosuchus* of some workers).

The most common fossils are teeth and plates (scutes) from the dermal armor. The teeth have bluntly conical crowns that are weakly but distinctly fluted. There may be a cutting edge or carina. The root is long and set in a deep socket. The scutes are rectangular plates of bone ornamented with deep, irregular pits on both sides. One side, the outer, may bear a ridge or keel.

Order Saurischia
some "dinosaurs"

While the term "dinosaur" has no standing taxonomically, it is too handy and too rooted in usage to be discarded completely. In proper use, it should be confined to terrestrial archosaurs of the orders Saurischia and Ornithischia. Unfortunately, popular usage is much less precise. Many of the large marine forms are included, as are the flying reptiles (Pterosauria).

Not all dinosaurs are large. The rumors of chicken-sized specimens are true. A few apparently were even tree-dwellers and obviously could not grow to large size. The two orders are distinguished mainly on the structure of the pelvis, but these criteria do not need to be treated here.

There are two recognized suborders of the Saurischia. The Theropoda ("beast-footed") includes all the carnivorous dinosaurs, from the tiny *Podokesaurus* (the infamous chicken-sized dinosaur) to such a ravening monster as *Tyrannosaurus* with its 4-foot skull armed with 6-inch teeth. The Sauropoda ("lizard-footed") comprise a group of huge, placid (we presume) herbivores such as *Brontosaurus*. These include the largest of the dinosaurs, with specimens at least 87 feet long, and perhaps exceeding 100 feet. The range of the order is Triassic to Cretaceous, and specimens are known from all continents except (as usual) Antarctica. No sauropods have been found in Alabama.

Suborder Theropoda

As noted above, this group includes the carnivorous dinosaurs. All forms are bipedal, and the jaws (except in one family) are armed with daggerlike teeth. Even a single tooth is diagnostic of the suborder. All are flattened, slightly curved daggers, ranging from tiny up to 6 inches in length, with both edges strongly serrated.

?FAMILY DEINODONTIDAE (fig. 69)

The family name means "terrible teeth." This designation is appropriate, for this family includes the huge *Tyrannosaurus* (which is only doubtfully distinguishable from *Deinodon*). The only material of this family (and order) yet known from Alabama is a single toe bone, reported by Langston (1960, p. 340, fig. 161), collected near Harrell Station, Dallas County. Langston identifies it as the proximal phalanx (basal toe bone) of the left middle toe. We regret to announce that it is not from a *Tyrannosaurus*. Rather, it comes from an animal less than half his size (we would estimate about 10 to 15 feet long). Still, he would be a bad customer in a dark alley. Langston (1960) notes that the bone most closely resembles *Gorgosaurus*, which is a smaller relative of *Tyrannosaurus* but still half the size of the largest of the genus. We estimate the teeth to be about 2 to 3 inches long and should appreciate hearing about any such finds.

Order Ornithischia
the rest of the "dinosaurs"

This is a complex assemblage of two- or four-footed plant eaters. They are united by the structure of the pelvis, which is quite different from that of the Saurischia. Four suborders are recognized.

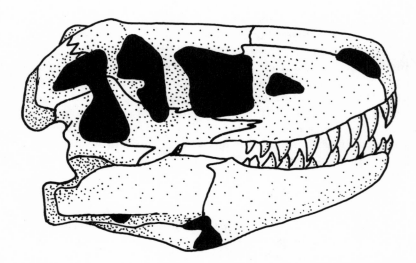

FIG. 69.   *Tyrannosaurus* (Deinodontidae). After Beerbower, 1968, fig. 20-8. Approximately 1/18 natural size.

The Ornithopoda ("bird-footed") is a bizarre group of bipedal reptiles. Some are comparatively ordinary in appearance. Others have greatly thickened skulls (the original "boneheads"), with the skull roof up to 8 inches thick, of solid bone, ornamented with knobs and spikes. Perhaps it was used as a battering-ram. Others have strange crests on the skull; these contain complex nasal passages and may have been air-storage reservoirs for amphibious forms, or resonating chambers for mating calls. Still others have the snout expanded into a ducklike beak or bill. This suborder includes the only really decent dinosaur specimen yet described from Alabama.

The Stegosauria include four-footed herbivores, all with spiked tails and bony plates along the back. The long-familiar restorations of *Stegosaurus,* with its double row of upstanding bony plates, may be faulty. Recent work indicates that the plates drooped down the sides, a less decorative but more functional arrangement. However, still more recent work puts them back up as heat controllers. Stegosauria remains are not known to occur in Alabama.

The Ankylosauria include solidly armored dinosaurs that manage to look like enormous armadillos (but are not related). The armor may bear strange spikes, particularly along the sides of the body. The tail may be spiked, or its last vertebrae enlarged and fused into a solid bludgeon of bone weighing up to 30 pounds. A whack on the shins from that would probably give the largest carnosaur second thoughts, if not a broken leg. Fragments are known from Alabama.

The Ceratopsia include the horned dinosaurs, with huge bony frills to protect the neck and a fearsome and varied arsenal of horns. They are known in rocks of Late Cretaceous age. These rocks are widely exposed in Alabama, but no ceratopsians have yet been reported.

Suborder Ornithopoda

Representatives of this order include both the oldest and some of the youngest of the Ornithischia. See above for a brief description. They are known in rocks from Triassic through very Late Cretaceous age.

FAMILY HADROSAURIDAE
duck-billed dinosaurs

Lophorhothon atopus Langston (fig. 70)

Lophorhothon atopus Langston, 1960, pp. 321–344, figs. 146–158, 159 (part), 163 (part).

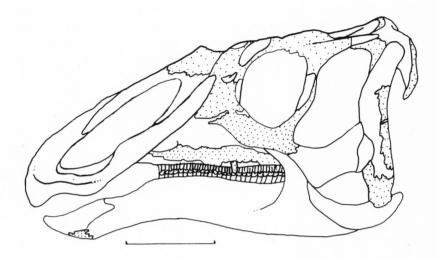

FIG. 70. *Lophorhothon atopus* Langston. Restored skull outline. Only the stippled parts are preserved; the rest is based on near relatives. That this is the *best* dinosaur skull known from Alabama to date suggests that we have much to learn. After Langston, 1960. Scale bar 10 cm.

This is the only really identifiable dinosaur yet (1980) found in Alabama. The specimen was found near Harrell Station, Dallas County. Lull and Wright (1942) earlier reported fragments from Dallas and Autauga counties, but these were unidentifiable.

The specimen consists of less than half of a skull, a series of vertebrae, and large parts of the fore and hind limbs. For a thorough description of the material, see Langston (1960). We can only hit the "high points" here.

The position of the bones and the particular bones preserved suggested to Langston (1960) that this specimen represents a dried carcass that was lying on the beach. It was then picked up by tides or waves and floated out to sea. The bones on the soft, rotting undersurface probably were left behind; others gradually dropped out through further decomposition. The carcass finally became waterlogged and sank to the bottom where it was slowly buried. This concept makes considerable sense; it would explain the very scattered occurrences of dinosaur bones with occasional partial skeletons that are found in marine rocks around the world.

The animal was relatively small, and the poor ossification of some of the bones is an indication that it was young. The total length was estimated at 15 feet.

The most likely finds of this species in the future would be isolated vertebrae and teeth. The teeth are about an inch long, roughly leaf-shaped, and are enameled only on the labial (outer) side. The edges of the enamel are thickened and regularly serrated. These teeth were part of a complex series of several rows that formed a broad grinding surface. The vertebral centra generally resemble those of plesiosaurs in that both ends are slightly concave in part of the series. Other vertebrae have a weak posterior concavity and anterior convexity. Langston (1960) notes that this feature is prevalent in the hadrosaurs but is less developed in young individuals like this specimen. Both ends of the vertebrae are strongly flared, and the center is somewhat constricted except on the dorsal surface (plesiosaur vertebrae are almost cylindrical). The neural arch (the bony covering over the spinal cord) is very tall.

This specimen should act as a spur to fossil hunters in Alabama. There is much to be learned.

Suborder Ankylosauria
armored dinosaurs

This group is described above. There are two recognized families, but only the Nodosauridae are known in North America. Langston (1960, p. 344–346, fig. 160, in part) reports a single partial ilium (part of the pelvis) from the Harrell Station area, Dallas County, Alabama, in the Mooreville Chalk. We will not figure this bone, as it would be meaningful only to a dinosaur specialist (and not particularly much to that person). There is also a single tooth from the Mooreville Chalk of Hale County.

The most likely finds of ankylosaurs are the plates that make up the armor. These are polygonal, rather thick plates of bone, roughly 1 to 3 inches across. A common ornamentation is a large central boss surrounded by one or more concentric rows of smaller beads. The undersurface commonly has a large, squarish projection for attachment into the skin.

Subclass Lepidosauria

See above, under Reptilia, for a description of this subclass and brief notes on the three orders included. Only one of these orders, the Squamata, is known in the fossil record of Alabama.

Order Squamata
lizards and snakes

Suborder Lacertilia
lizards

The lizards are among the most familiar and flourishing of the living reptiles. Their place in the ecology of North America is important, but often overlooked. They are among our principal allies in the continual war against insects.

Most lizards, of course, have four legs. A few legless forms are often confused with the snakes. These are the glass snakes, so called because they may break if handled. They retain the ability of many of their more normal cousins to shed the tail to foil an attack, so the still writhing tail perhaps holds the attention of the attacker long enough for the truncated lizard or glass snake to make a getaway. The lizard can then grow a new tail—not so long and fine as the original, but equally useful and readily shed. The glass snakes can be distinguished from the true snakes by their belly scales. In the snakes, each belly scale extends from side to side across the ventral surface. The belly of a glass snake is covered with a mosaic of small scales.

Only one family of lizards is known as fossils in Alabama. They bear very little resemblance to the shy creatures that sun themselves on your back fence.

FAMILY MOSASAURIDAE
mosasaurs

These are moderate-sized to gigantic marine lizards, found only in rocks of Late Cretaceous age. A small skeleton would be about 6 feet long; the largest (Thurmond, 1969) may have reached a size of 55 feet, with a skull as long as the entire body of a small mosasaur. All of the mosasaurs, except for two peculiar genera, were active and ferocious predators, feeding largely on fishes but probably not disdaining smaller reptiles. The two peculiar forms had teeth specialized for cracking and eating shellfish.

Mosasaur vertebrae are extremely common in the Mooreville Chalk of Alabama (Russell, 1970). More complete specimens are found occasionally. The vertebrae are highly distinctive; each vertebral centrum is procoelous, that is, it has a deep cup at the front end and a matching ball at the posterior end that fits into the cup of the next vertebra. The vertebrae from different regions of the body also are easily distinguished. The cervical (neck) vertebrae are about as long as tall (centrum only) and have a single large facet on the ventral surface, directed down and back. This facet is for an extra knob of bone below the centrum, termed the intercentrum. On posterior cervicals, the intercentrum is small and may be fused to the centrum. There usually are seven cervicals, but the first (called the atlas) is very unusual in form and hardly recognizable as a vertebra.

The dorsal (trunk) vertebrae normally are considerably longer than wide or high. There are no ventral facets. The sacral (hip) vertebrae are hardly distinguishable from the dorsals, because the pelvis is not attached to the vertebral column in mosasaurs. The caudal vertebrae are shorter than high. At the base of the tail these vertebrae bear two long transverse processes, one on each side, directed laterally and slightly downward, but there are no ventral facets. After a few tail vertebrae, a pair of articular facets appears between the bases of the transverse processes. These support the chevron bones, Y-shaped coverings for the main blood vessel of the tail. In some mosasaurs the chevron bones are fused to the vertebrae rather than articulated. The tail vertebrae gradually become smaller toward the posterior and become less well defined. The last few are little more than nondescript modules of bone.

The skull and jaws bear features important in identifying mosasaurs and need to be discussed in some detail. The most anterior bone of the skull is the premaxilla, which forms the tip of the snout and bears the first two teeth on each side. The premaxilla sends a long median process backward along the top of the snout to form a bony bar between the nostrils. The shape of the tip of the premaxilla is particularly important in identifying mosasaurs; a feature to note is whether the tip of this bone lies directly above the roots of the first teeth or is prolonged into a bony prow called a rostrum.

The main tooth-bearing element of the upper jaw is the maxilla. There are twelve to eighteen teeth in the maxilla. The number of teeth may help in identification but is tricky to use, as it may vary within a single species.

The palate or roof of the mouth is composed of a complex series of bones, most of which do not need discussion here. The most important bone is the pterygoid, a paired bone at the back of the palate. This bone is roughly in the shape of a very twisted X, one of whose bars is long and the other short. It bears a series of teeth (seven to fifteen), which are usually much smaller than the jaw teeth and are more strongly curved. They served to hold prey while it was being swallowed, especially if the prey were large. Most snakes have similar teeth on the pterygoids.

The braincase is complex and will not be discussed here, though it

is of great technical importance in determining relationships among
the mosasaurs. Basically, we are ignoring it to prevent this discussion
from becoming too much like a course in comparative anatomy.

The roof of the skull is composed of a series of large platelike bones
(fig. 71). We have already mentioned the maxillae and the premaxillae
that form the anterior portion. Between the maxillae and the eye sockets (technically called orbits) are the prefrontals, usually roughly
diamond-shaped. The main bone of the skull roof is the frontal, a large
median element that is basically wedge-shaped, with the tip of the
wedge cut away to receive the posterior end of the premaxilla. Behind
the frontal is the complex parietal, which covers the main part of the
brain. In top view, this bone is a Y-shaped element consisting of a base
that is the braincase roof and two long wings extending out and back
toward the jaw joint. The base of the Y is expanded and bears a distinct
hole, the parietal foramen (also called pineal foramen), at or near its
contact with the frontal. The parietal may or may not have a broad, flat
table above the braincase, a feature to look for. The upper temporal
openings are large and lie to either side of the parietal. Their outer
boundary is composed of two bones, the postorbitofrontal and the
jugal. The postorbitofrontal, called the postorbital in most older work,
is a strange four-branched bone. One branch extends forward beside
the frontal to become the top of the orbit (eye socket). Another runs
straight down the back of the orbit to meet the jugal, an L-shaped bone
that forms the lower border of the eye socket. The third branch heads
off across the back of the frontal to meet the parietal. The fourth
branch is the largest; it goes straight back to constitute much of the bar
between the upper and lower temporal openings and to meet the
squamosal, which forms part of the jaw joint and connects the postorbitofrontal with the parietal.

One of the strangest bones of the mosasaur skull is the quadrate, a
pulley-shaped bone that connects the lower jaw to the skull. It was
movable at both ends and allowed the lower jaw to move forward and
backward to a considerable degree. The outer surface of this bone
bears a large, smooth frill that may show traces on its surface of the
eardrum (tympanum), reinforced with calcium deposits in many
mosasaurs and often preserved in a fossil. The quadrate forms a broken ring (complete in one genus) around a central opening. The stapes
bone, a slender sound transmitter between the eardrum and the skull,
ran through this opening.

The lower jaw is extremely complex, but its structure may well be
the key to the rapid (though brief) success of the mosasaurs. The
quadrate allowed a fore-and-aft motion for the lower jaw. The rest of
the jaw is made up of six bones. The only tooth-bearing element is the
dentary, with eleven to eighteen teeth. The tip of the dentary may be

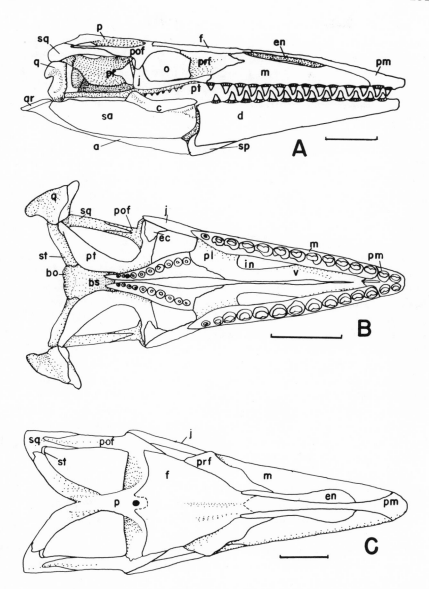

FIG. 71.   Mosasaurian skulls. A: *Tylosaurus*, lateral view. B: *Platecarpus*, ventral
view. C: *Prognathodon*, dorsal view. After Russell, 1967. Scales differ; all
scale bars 10 cm.

a, angular; *ar*, articular; *bo*, basioccipital; *bs*, basisphenoid; *c*, coronoid; *d*,
dentary; *ec*, ectopterygoid; *en*, external nares; *f*, frontal; *in*, internal nares; *j*,
jugal; *m*, maxilla; *o*, orbit; *p*, parietal; *pl*, palatine; *pm*, premaxilla; *pof*, postorbitofrontal; *pr*, preotic; *prf*, prefrontal; *pt*, pterygoid; *sa*, surangular; *sp*,
splenial; *sq*, squamosal; *st*, supratemporal.

drawn out into a toothless counterpart of the premaxillary rostrum; it is
important to watch for this character. Ventral to the dentary, and forming most of the lower border of the jaw, is the splenial. Through most
of its length, this bone is a trough in which the dentary rests. It extends much farther up the lingual side of the jaw than does the labial.
At its posterior end is a socket into which fits a complex ball on the
back part of the jaw. Oddly enough, a hinge at midjaw, behind the
dentary and splenial, allowed limited movement. The lower jaws
probably could have closed in the manner of a set of parallel-jawed
pliers, rather than in a scissorslike action. This method of closing
could be handy if a fish were fleeing just at the front of the jaws. The
anterior ends of the lower jaws were not fused but held together
loosely by ligaments. This lower jaw is much like that of a snake,
which also has the midjaw hinge.

The posterior part of the lower jaw, behind the midjaw hinge, is
composed of four bones. The largest is the angular, which forms the
entire ventral margin. It extends from a complex ball that fits into the
splenial (fig. 71A) to a stubby process that goes beyond the jaw joint to
provide attachment for the muscles that opened the jaw. Resting on
the angular near its posterior end (and sometimes fused to it) is the
articular, which receives the lower end of the quadrate. In front of the
articular and above the angular is the massive platelike surangular.
Just behind the midjaw joint, a saddle-shaped coronoid rests on the
surangular to provide the main attachment for the muscles that close
the jaw. In side view, the top of the coronoid may be either nearly flat
or deeply curved, a distinguishing feature.

The bones of the limbs also are distinctive, but classification
requires a discerning eye and, usually, an identified comparative collection from which to work. The phalanges (finger and toe bones)
normally are much longer and more slender than those of plesiosaurs
and are roughly hourglass-shaped.

This rather lengthy introduction to mosasaurs has been necessary to
prevent repeating, *ad nauseum*, these details in the discussions of the
individual species. It is also needed because the skulls of mosasaurs
are very poorly knit together (the whole skull seems to have been
capable of much internal movement), and isolated skull bones are
often found.

The relationships of the mosasaurs to other reptiles seem to have
been rather thoroughly determined. Oddly enough, they seem to be
descended from a family of lizards that, though ancient, is still
living—the Varanidae, which includes the monitor lizards of Africa,
Asia, and Australia. They are all predatory and many are semiaquatic.
Some still grow to very large sizes. The largest of living lizards is the
Komodo dragon, *Varanus komodoensis*, native to a few small Indone-

sian islands. These lizards attain a length of at least 9 feet (12-footers are rumored) and weights in excess of 200 pounds. Though feeding mostly on carrion, they are reported to be fast enough to run down deer and powerful enough to kill wild boar.

The mosasaurs swam mainly by fishlike lateral undulations of the body, as do the living monitors, using the paddles for steering and braking. The structure of the vertebral column, with its many ball joints with very smoothly finished surfaces, indicates considerable flexibility. In this connection, it is interesting to note that many specimens show signs of arthritis, often to the extent of fusing several vertebrae. There is something pathetic in the thought of an ancient, arthritic monster creaking its way through a Late Cretaceous sea until starvation or disease finally ends its sufferings. The arthritic condition also indicates, though, that once the perilous days of youth were past, many lived to a ripe old age.

The mosasaurs of Alabama are a very varied group. Fortunately, two excellent recent studies are available. Russell (1967) reviewed the entire North American mosasaur fauna in considerable detail, with numerous well-executed and useful figures. Later, Russell (1970) devoted considerable work to Alabama material. Both these works are still in print and are reasonably priced. They should be acquired by anyone intending serious study, as they provide far more detail than we can here. In the older literature, Cope (1875E) and Williston (1898) have many excellent figures, both of individual bones and of restorations, mainly of the famous Kansas material.

Russell's work has not exhausted the possibilities of the mosasaur fauna of Alabama. Many forms remain very poorly known; the authors are aware of at least two specimens that have not been reported so far from Alabama. Work on these is not yet (1980) in a sufficiently advanced state for publication other than a brief note.

Subfamily Mosasaurinae

The premaxilla bears a short rostrum. The dentary and maxilla have at least fourteen teeth. Dorsal and cervical vertebrae in end view have the centrum shaped as a circle or flattened oval; the chevron bones are fused to the vertebral centra in the tail. The coronoid bone of the lower jaw is deeply curved and rises above the surangular in a tall process.

Clidastes propython Cope (fig. 72)

Clidastes propython Cope, 1869A, p. 258.

Clidastes propython, Russell, 1967, pp. 128–131, figs. 74, 75a, 76a; Russell, 1970, pp. 371–373, fig. 166B.
Many synonyms; see Russell, 1967, for listing.

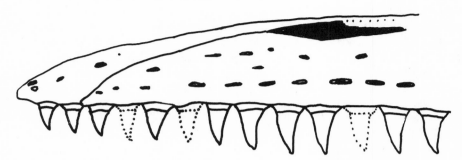

FIG. 72.   *Clidastes propython* Cope. Lateral view of anterior part of skull, based on a Kansas specimen. Note gradual slope of suture between premaxilla (snout) and maxilla (upper jaw). The anterior part of this suture is nearly vertical in *C. liodontus* (not yet definitely reported from Alabama). In the latter species, the external nares (nose openings) begin more anteriorly, about over maxillary teeth 4-6. After Russell, 1967. x½.

This species is the smallest and most common of the mosasaurs from the Mooreville Chalk. In most cases, its abundance in collections does not coincide with its true abundance because of its small size. In the extremely thorough collections by the Field Museum of Natural History, in which every bone fragment was collected (Zangerl, 1948A), this species comprises about 75 percent of the mosasaurian remains gathered (Russell, 1970). Russell notes considerable variation in size. The smallest of the Alabama specimens was estimated to be about 7½ feet long and the largest 20½ feet (Russell, 1970).

In many respects, the genus *Clidastes* is the most primitive of the known mosasaurs. The teeth have smooth enamel surfaces and cutting edges fore and aft. The vertebrae preserve an extra articulation (zygosphene-zygantrum), immediately above the neural canal, that does not appear in functional form in most other mosasaurs. There are sixteen to eighteen teeth in the maxilla and seventeen to eighteen in the dentary. The parietal bone has a broad, nearly flat dorsal table. The posterior hook of the quadrate (technically, the suprastapedial process) is long.

As far as is known, *C. propython* is distinguished from other members of the genus by several characters. The suture between the premaxilla and maxilla rises in a gentle curve, rather than going through a sharp angle, and the root of the last premaxillary tooth is exposed on this sutural surface. The anterior end of the parietal bears a pair of deep notches for processes of the frontal. The quadrate has a small process rising from its base toward the end of the suprastapedial process.

Isolated vertebrae, unless the neural arch is well preserved, showing the zygosphene-zygantrum, cannot be definitely assigned to species of *Clidastes*, nor even with certainty to the genus. Most specimens show an oval end view, widest from side to side. There are some circular vertebrae, which cannot be distinguished from those of *Mosasaurus* unless the neural arch is preserved. However, any specimen less than 2 inches across probably belongs to *Clidastes* or to a very young *Mosasaurus*.

The type specimen was collected from the "Rotten Limestone" (an old name for the Selma Group) near Uniontown, Alabama, by Dr. E. R. Showalter of Howard College. It was given to the Academy of Natural Sciences of Philadelphia, where it is still in the collections (#10193). To this day, it is the most complete mosasaur skeleton collected in Alabama. The species occurs throughout much of the Black Belt.

Halisaurus sternbergi (Wiman) (fig. 73)

Clidastes sternbergi Wiman, 1920, p. 13, figs. 4–9, pl. 3, pl. 4, fig. 5a–b.
Clidastes sternbergi, Russell, 1967, pp. 126–127, figs. 71, 75C.
Halisaurus sternbergi, Russell, 1970, pp. 369–371, figs. 164–165.

FIG. 73.   *Halisaurus sternbergi* (Wiman). Lateral view of anterior part of skull, based on a Kansas specimen. Only the peculiar, very flattened cervical vertebrae are known from Alabama as yet. After Russell, 1967. x⅔.

This form is poorly known. The vertebrae are very flattened, at least in the neck and forepart of the trunk, with articulations about twice as wide as high. The premaxilla has the rostrum reduced almost to the vanishing point. In side view, the snout is very low, and the suture between the maxilla and premaxilla rises at first almost vertically, goes through a sharp bend, becomes almost straight for a considerable distance, bends again, and runs almost straight to the front of the nostril. The nostril begins very far posteriorly, over the eleventh maxillary tooth, rather than over the seventh as in *Clidastes propython*. The suture between the frontal and the parietal is almost straight.

The type specimen is from Kansas. Russell (1970) reports this peculiar form from the Mooreville Chalk of Greene and Dallas counties, Alabama.

Caution: before referring a vertebra to this species, make sure it has not been crushed.

Globidens alabamaensis Gilmore (figs. 74A, 74B)

Globidens alabamaensis Gilmore, 1912, p. 479, figs. 1–3, pls. 39–40.
Globidens alabamaensis, Russell, 1967, pp. 144–145; Russell, 1970, p. 373.

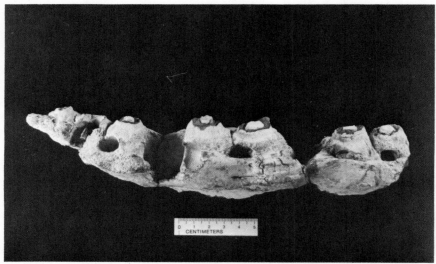

FIG. 74A. *Globidens alabamaensis* Gilmore. Lingual view of right dentary. Specimen in collection of University of Alabama Museum of Natural History is from the Mooreville Chalk near Harrell Station, Dallas County, Alabama.

This is the strangest of the Alabama mosasaurs. Until the discovery of a specimen by a University of Alabama archaeologist in 1977, the best specimen known was the type (Gilmore, 1912), consisting of a maxilla, other skull fragments, and a vertebra. C. B. Curren found the most recent specimen in the Bogue Chitto Prairies near Hamburg, Perry County, the locale of the type. Curren's find, in the collections of the University of Alabama Museum of Natural History, consists of most of the right dentary with eight to ten teeth, parts of both surangulars, parts of both splenials, part of at least one angular, three vertebrae, and miscellaneous skull and jaw fragments. At the time of this writing, attempts are being made to find more of this specimen, which may prove to be the best one known to date.

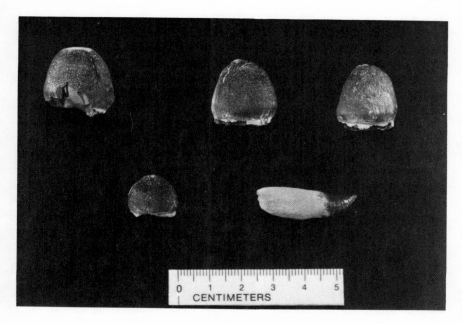

FIG. 74B. *Globidens alabamaensis* Gilmore. Lateral view of teeth associated with dentary and other fragments of the Harrell Station specimen. Globular teeth do not fit the dentary shown in Fig. 74A. Tooth at lower right may be from premaxillary or distal part of dentary.

The teeth are unique. Up to an inch across, they are exactly what the generic name implies—almost globe-shaped. There is a small hump at the very top of each tooth, and the enamel is ornamented with many grooves and ridges, running from the apex down toward the base.

Recent work perhaps has more confused than clarified our picture of this genus. The skull fragments associated with the type were quite thick; consequently, this genus has become a dumping ground for almost any unusually thick, unidentifiable scrap of skull. At least one other form of mosasaur in Alabama has an unusually massive skull, but it is as yet undescribed. The teeth of this skull are not at all like those of *Globidens* but are pointed in normal mosasaurian fashion.

Teeth of this species have been found also in Texas (Russell, 1967; Thurmond, 1969), Mississippi (Gilmore, 1927), and Belgium (Dollo, 1924). Related forms are known from Israel, Egypt, and the East Indies (Timor). A complete skull was described from South Dakota by Russell (in press). Although referred to a different species, *G. dakotaensis*, it differs little from the Alabama form.

Mosasaurus sp.

A private collector has a series of vertebrae, collected near Sawyer-ville, Hale County, Alabama, that probably belongs to this genus. The articular faces are nearly circular and over 3 inches across. We can say little more about this find at present.

Subfamily Plioplatecarpinae

The rostrum is absent, and the number of teeth in the maxilla and dentary is fairly small (twelve or more) compared to the mosasaurines. The coronoid bone is small and nearly flat, except in *Prognathodon*. The vertebral centra are almost all roughly pear-shaped in end view. The chevron bones are articulated to the caudal centra, not fused.

Two genera have been reported from the Mooreville Chalk of Ala-bama, *Platecarpus* and *Prognathodon*. Like the other mosasaurs, the plioplatecarpines are found only in rocks of Late Cretaceous age.

Platecarpus sp. (fig. 71B)

This genus is composed of medium-sized mosasaurs (about 20 feet in length) that are more stoutly built than are those of the genus *Clidastes*. The skull and jaws are more massively proportioned and relatively shorter than those of mosasaurines. For some odd reason, this genus is much rarer in Alabama than in rocks of equivalent age in Kansas and Texas. The large Chicago collections from the Mooreville Chalk contain only a single specimen of *Platecarpus*, while this genus makes up about three-fourths of the Kansas mosasaur material (Rus-sell, 1970) and half of that collected in Texas (Thurmond, 1969). This rarity in the Chicago collections at the Field Museum may be partly accidental; bones of *Platecarpus* are much more common in the col-lections of the University of Alabama Museum of Natural History and the Geological Survey of Alabama. However, both of these collections tend to concentrate on the larger and heavier bones, giving a bias toward *Platecarpus* and away from *Clidastes*.

Platecarpus has two teeth in each side of the premaxilla. The maxilla bears twelve teeth, and the dentary eleven or twelve. The posterior hook of the quadrate is extremely large and broad, and the quadrate forms nearly a complete ring. The coronoid bone of the lower jaw is small and almost flat.

To date (1980), no available specimen of *Platecarpus* from Alabama has been referred to a definite species. A reexamination of the skele-ton described and figured by Dowling (1941) should provide an iden-tification, but the location of the specimen is unknown to the authors.

There seem to be five recognizable species of *Platecarpus* thus far known from the Cretaceous of North America. Russell (1967) notes four of them, and Thurmond (1969) resurrects an old name rejected by Russell (1967). We will briefly state the characters by which each can be recognized.

Platecarpus tympaniticus Cope, 1869A, p. 265.

This poorly known species is represented only by a fragmentary specimen from the Eutaw Formation of Mississippi, near Columbus. It is the type species of the genus *Platecarpus* and probably is a senior synonym of either *P. coryphaeus* or *P. ictericus*, but this cannot be determined until more material has been found. Thurmond (1969) suggests, on stratigraphic grounds, that it is probably *P. coryphaeus* and would thus replace that name.

Platecarpus coryphaeus (Cope, as *Holcodus coryphaeus*, 1872D, p. 269).

As noted above, this species perhaps should be called *P. tympaniticus*. The suture between the maxilla and the premaxilla ends at the front of the nostril, above the third tooth in the maxilla. On the outer face of the anterior end of the dentary the openings for a branch of the trigeminal nerve separate into two rows at a point below the fourth tooth, the two rows rejoining at the tip of the dentary.

Platecarpus ictericus (Cope, 1871, p. 572, as *Liodon ictericus*).

This species is about the same size as *P. coryphaeus*, with a dentary about a foot long. The differences between *P. ictericus* and *P. coryphaeus* are very subtle. In *P. ictericus* the suture between the premaxilla and the maxilla ends (and the nostril begins) between the second and third teeth in the maxilla. The two rows of openings for a branch of the trigeminal nerve on the outer surface of the dentary run to the tip of the bone without separating. Russell (1970), on stratigraphic grounds, suggests that the Mooreville *Platecarpus* most likely belongs to this species. A specimen in the University of Alabama Museum of Natural History (number V-4), collected from about 30 feet above the base of the Mooreville Chalk in Greene County, Alabama, probably belongs to this species, as the nostril begins between the second and third teeth on the maxilla.

Platecarpus curtirostris (Cope, 1872D, p. 278, as *Liodon curtirostris*).

Russell (1967, p. 154–155) placed this species in synonymy with *P. ictericus*, because the jaw structures, particularly the openings for the trigeminal nerve, are similar. Thurmond (1972) considers this species distinct on two characters. The quadrate forms a complete ring, though not fused (there is a gap in *P. ictericus*). The posterolateral branches of the parietal are concave forward, rather than convex, giving room for a more powerful jaw musculature. It may show tendencies leading toward *Prognathodon*. There are no reports of specimens from Alabama, nor are any specimens from Alabama known to the authors.

Platecarpus sp. cf. *P. somenensis* Thevenin (1896, p. 907).

Originally, this species was described from northern France. Russell (1967) reports similar material from South Dakota, and Thurmond (1969) from Texas. It is much larger than any of the preceding species of *Platecarpus*, the dentary being 1½ to nearly 2 feet long. The teeth are striated rather than faceted. Again, there are no indications of this species from Alabama.

Except for this species, all known species of *Platecarpus* have strongly faceted teeth. All species have teeth with powerful cutting edges fore and aft.

"Platecarpus" intermedius (Leidy)

Clidastes intermedius Leidy, 1870, p. 4.
Clidastes intermedius, Leidy, 1873B, pp. 281, 344, pl. 34, figs. 1–2, 4–5, 10.
"Platecarpus" intermedius, Russell, 1967, p. 156; Russell (1970), pp. 369, 378.

Only a single fragmentary specimen of this poorly known species has been found. This specimen is now in the collections of the Academy of Natural Sciences of Philadelphia under the numbers 9023, 9024, 9029, and 9092–9094. It was collected in Pickens County, Alabama, from the Selma Group ("Rotten Limestone") by Dr. J. C. Nott, who had resided in Mobile.

The vertebrae have functional zygosphenes, the characteristic by which Leidy (1870, 1873B) originally referred the species to *Clidastes*. Russell (1967) considers the jaw to be more like that of *Platecarpus* but thinks this generic assignment tenuous at best. It may represent a new genus when better material is available. The teeth, at least at the posterior end of the dentary, are peculiarly short and inflated, indicating a diet of shellfish like that of *Globidens*.

Prognathodon sp. (fig. 71C)

Russell (1970) has the first report of this genus from the Mooreville Chalk of Alabama. There are several fragmentary specimens from Greene and Dallas counties. The Selma *Prognathodon* is too poorly known at present to be referable to any of the known species, but Russell (1970) notes that the material more closely resembles *P. solvayi* from Belgium than it does any of the previously known American species.

The skull is extremely heavily built. The teeth are not swollen but are quite stout, rather blunt cones, numbering twelve in the maxilla and fourteen in the dentary. They bear two prominent cutting edges, and the enamel surface is smooth in most American species. However, it is faceted in the Alabama material and in *P. solvayi*. The quadrate is unique; it is solidly fused into a complete ring.

Our restoration of the skull is from Russell and is based on *P. overtoni*. The Alabama material differs in two respects from this restoration, according to Russell (1970). The tip of the dentary is very abruptly terminated, with the first tooth overhanging the end of the jaw. The jugal is much more massive and has a prominent process directed posteriorly. Such a process is all but lacking in both *P. overtoni* and *P. solvayi*. If its presence proves consistent in the Alabama material, a new species name might be appropriate.

The teeth of *Prognathodon* appear adapted for eating a combination of larger fishes (plus some small reptiles) and thin-shelled mollusks. In this connection it is interesting to note a paper by Kauffman and Kesling (1960) on a large ammonite (a cephalopod; see Copeland, 1963, p. 29) showing tooth marks of a mosasaur. Careful study of the bites indicated that the assailant was a *Prognathodon* and gave interesting details of the *modus operandi* of this Cretaceous mugger.

Subfamily Tylosaurinae

The tip of both premaxilla and dentary are prolonged into a prominent rostrum. The teeth are large with sharp cutting edges that may be serrated in very large specimens (Thurmond, 1969). The vertebrae are pear-shaped in anterior view, but the articulations are not so prominent as in most other mosasaurs. The posterior hook of the quadrate is short, and the coronoid bone is nearly flat, though not so reduced as in plioplatecarpines. The chevron bones are not fused to the caudal centra. The only eastern North American genus known is *Tylosaurus*, which ranges in size from merely large to gigantic. Thurmond (1969) reports a skull from Texas over 6 feet in length.

Tylosaurus is reported from Alabama by Renger (1935A and 1935B as *Mosasaurus*), and by Russell (1970). The material known is very scrappy. The most common North American forms are *T. proriger* and *T. nepaeolicus*. The Alabama material seems to belong to neither of these species.

Tylosaurus zangerli Russell (see fig. 71A, another species)

Tylosaurus zangerli Russell, 1970, pp. 374–375, figs. 169–170.
?*Tylosaurus* sp., Russell, 1970, pp. 375–376, fig. 171.

It is difficult to compare *Tylosaurus zangerli* to the other species of *Tylosaurus*, as it is known only from a humerus and a femur. These are certainly distinctive, being much more slender in the middle than these bones in a normal *Tylosaurus*, and with the ends much more squarely cut. The animal must have been fairly small.

For convenience more than for any taxonomic reason, we associate here the fragmentary skeleton assigned by Russell (1970) to *Tylosaurus* sp. The most distinctive feature of this material is on the inner side of the lower jaw at the midjaw joint. There is a deep notch on the angular into which fits a corresponding process on the splenial. This deep notch also occurs on larger specimens (particularly V-7, from Perry County, Alabama) in the collection of the University of Alabama Museum of Natural History but is lacking on specimens from Hale and Lowndes counties (exact localities unknown) in the collections of the Department of Geology, Birmingham-Southern College. It may be an individual variation, or it may represent the presence of two species of *Tylosaurus* in the Mooreville Chalk.

The long rostrum of *Tylosaurus* is its most distinctive feature. It is often as long as the tooth-bearing part of the premaxilla. It is possible that the rostrum was used as a battering-ram, to stun enemies or prey. At least one specimen (uncatalogued) at the University of Kansas Museum of Natural History shows considerable "mushrooming" of the rostrum that was subsequently healed (almost a "cauliflower nose"). Either it was used as a ram or we know of one extremely clumsy tylosaur.

Suborder Serpentes
snakes

The snakes should need no further descriptions! See above for distinctions from the legless lizards. They seem to be descended from lizards not unlike the living monitors (or the mosasaurs). At least, both have the midjaw joint, toothed pterygoid bones in the back of the throat, and forked tongues. The earliest known snakes come from Early Cretaceous deposits.

FAMILY PALAEOPHIDAE
primitive marine snakes

This family is the oldest of the snakes and is known mainly from nearshore marine deposits. They may have been either aquatic or terrestrial; there has not been enough work on nearshore terrestrial deposits to know whether these marine snakes existed in that environment. The paleophids range from Early Cretaceous through the Eocene. The single Alabama specimen is one of the latest known.

Pterosphenus schucherti Lucas (fig. 75)

Pterosphenus schucherti Lucas, 1898B, p. 637, pls. 45, 46.

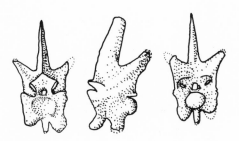

FIG. 75.   *Pterosphenus schucherti* Lucas. Thoracic vertebra, anterior, lateral, and posterior views. After Lucas. x½.

It is not necessary here to describe this species in detail. The specimen was collected in the Jackson Group (upper Eocene) at Old Cocoa Post Office, Choctaw County, Alabama.

In general, snake vertebrae look rather like miniature mosasaurs, with an anterior cup and a posterior ball. The processes are much larger relative to the centrum, and extra articulations between the centra are present.

Class Aves
birds

It is difficult to mistake a living bird for any other vertebrate. The only question to ask is, "Does it have feathers?" All birds are feathered; no other vertebrate (except for a few human malefactors who were given a preliminary "skin" of tar) has them.

Paleoornithology, the study of fossil birds, is the rarest and most esoteric of the divisions of vertebrate paleontology. Bird bones are usually small and always fragile. Their walls are very thin, and the marrow cavity often contains an air sac for further lightness. This fragile structure makes them the rarest of vertebrate fossils.

Bird bones are also extremely difficult for any but a specialist's specialist to identify. Though ranked as a class containing numerous orders, skeletal differences between very distinct birds are often elusive. Identification requires a magnificent collection of Recent skeletons such as is found in only a few of the largest museums.

Only two fossil birds have ever been reported from Alabama, and only one of them is authentic. The other, *Alabamornis gigantea* Abel, 1906, was later (Kellogg, 1936A, p. 7) shown to be based on a pelvis of a primitive whale *(Basilosaurus)*.

Order Ciconiiformes
herons, storks, flamingoes

Suborder Ciconiae
storks and ibises

FAMILY PLEGADORNITHIDAE

Plegadornis antecessor Wetmore (fig. 76)

Plegadornis antecessor Wetmore, 1962, pp. 1–3, fig. 1.

This bird is based on a single humerus from the Mooreville Chalk of Greene County. No other material as yet is known. It is the only representative of its family and the oldest known member of its suborder.

Fossil bird material of this age is extremely important. If you believe that you have found such, contact professional help immediately.

Class Mammalia
mammals

At present the mammals are the most successful of the vertebrates and seem to represent their most efficient development to date. They are, and have been since the close of the Cretaceous, the dominant

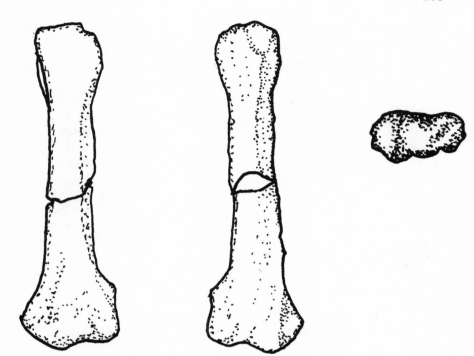

FIG. 76. *Plegadornis antecessor* Wetmore. Type, humerus. About twice natural size.

vertebrates on land and have invaded the air and the sea with considerable success.

There are two keys to this success. The first is the development of warm-bloodedness, the regulation of the internal temperature of the body. The general idea of this warm-bloodedness is the maintenance of all the organs of the body at their most efficient temperature for functioning, regardless of outside conditions. Many features of mammals are associated with the maintenance of internal temperature. The development of a coat of hair serves as an insulating blanket around the body; only mammals have hair. The hair traps a layer of still air next to the skin, providing this insulation. In water, such a protection does not serve during long periods of submergence. The hair in most aquatic mammals is replaced by an insulating blanket of fat (blubber) under the skin. Even so, all aquatic mammals retain, at least at some stage of development, their ancestral fur coats.

Also associated with the maintenance of temperature is the pose of the mammalian body. Reptiles tend to have sprawling gaits; the belly rests on the ground during stops and often is dragged while walking. This contact with the ground allows enormous heat loss. In mammals, the legs are brought directly under the body, and the belly is off the ground during stops as well as when moving. A few forms, like the horses, even sleep standing. A side effect of this shifting of the legs to a position under the body is more efficient locomotion. A large part of a reptile's walking effort is devoted to merely lifting the belly off the ground—many reptiles don't even bother with this. Much less energy is available for the actual job—moving from Point A to Point B. Almost all of a mammal's walking energy gets the body somewhere.

The second key to mammalian success is the care of the young. After birth, most young reptiles are strictly on their own. A few forms, such as the alligators, may help the young out of the nest. That is the limit of reptilian maternal care. A mammalian mother has the equipment, both physical and habitual, to care for the young for a significant time after birth. The physical equipment consists of a means of secreting milk to feed the young. The habitual equipment consists largely of a built-in need to protect and educate the baby mammal (a hatchling reptile is usually fair game for anyone, even Mama, except among alligators). Thus, a young mammal finally emerges to face the world in a far more developed state than does any young reptile.

The reptile-mammal transition is the best documented passage between classes in the entire fossil record. Fossil forms exhibiting both reptilian and mammalian characteristics are known (*Eozostrodon,* or *Morganucodon,* for example). Living curiosities are the monotremes—the duckbill platypus and the echidna and spiny anteater. These animals have body hair and mammary glands but lay eggs.

Other primitive "transitional" forms are the marsupials (opossum, kangaroo, etc.). These give birth to only partially developed offspring, which are transferred to the mother's marsupium, or pouch, where they attach to milk glands and essentially are force-fed until fully developed.

The characteristics exhibited by these more primitive mammals or mammallike creatures do not permit a high degree of competition with more advanced forms. Survival of any plant or animal group depends on the reproductive success of the species. A group that cannot compete successfully for space and food cannot keep reproduction at a replacement level. Adaptability is the key to survival; the mammals display the highest degree of adaptability of any vertebrates.

Other groups of vertebrates have attained partial solutions to the problems solved by the mammals. The birds in particular came remarkably close. They are warm-blooded; they have insulating coats (feathers rather than hair). In the Early Tertiary, following the rather mysterious demise of the large reptiles at the close of the Cretaceous, the issue of who would inherit the earth seems to have been in doubt for a time. There were large and powerful ground-living birds, some standing over six feet tall, while the contemporary horse was the size of a fox terrier. The mammals, though, had an edge, or several edges. Almost all mammals bear the young alive; there is no vulnerable (and tasty!) egg stage to be faced. Probably the greatest advantage the mammals had was the presence of four highly adaptable feet. The birds had gained their advantages at the price of specializing the forelimbs into wings that had lost all possibility of developing into anything else (except swimming flippers, as in the penguins). In those birds, such as the ostrich, that became purely ground-dwelling, the wings are reduced into almost useless appendages.

A mammal's forelegs had infinitely more possibilities. They could become grippers or hooks for tree climbing, powerful thrusters for running, spades for digging, fearsome weapons for attack and defense, even wings as in the bats. The mammalian pattern was adaptable for a vastly more varied series of life styles than was the bird pattern. Another price the birds paid for their success was the accident that they began with primitive teeth—something that all living birds have dispensed with entirely. Mammalian teeth came from highly developed and adaptable ancestors and provided almost infinite possibilities. They have become grinders for plant materials or daggers and slicers for meat; only a very few mammals have seen fit to dispense with these useful organs. As a result, the teeth of mammals are the most important parts for determining relationships. They are durable and easily fossilized, they carry a valuable record of life habits, and they have been intensely studied from almost every angle. Many mammals are known only from their teeth.

The patterns of mammal teeth are as varied as the mammals themselves. Of particular importance are the back teeth (molars). These come in many patterns. A key to the mammalian character of a tooth is the presence of several points (cusps) on a single tooth, arranged in a fairly complex pattern. We will not go into the complex terminology that has been developed for naming cusps; suffice it to say that almost every crinkle and point of a mammal tooth has a specific name.

However, we can trace the general development of the teeth, and the names for some of the more prominent and important patterns. The teeth of primitive mammals seem to be basically triangular, with varied arrangements of sharp cusps. These animals invariably are small, and the teeth seem to be associated with insect-eating habits. The sharp cusps serve to penetrate and break up the tough shells of most insects; the bodies of the prey can then be further ground in the spaces between the cusps so that they will digest more rapidly and efficiently. From this early pattern, the teeth developed in various directions.

If the animal becomes specialized for plant eating, the cusps tend to become blunted and widened to provide a broader grinding surface. An early type is the *bunodont* tooth; the cusps, while either retaining their primitive pattern or becoming squared up by addition (on uppers) or loss (on lowers) of cusps, become blunted into low mounds. "Bunodont" literally means "hill-tooth." The bunodont pattern is associated with an omnivorous diet; the animal is equipped to eat almost anything, though not with the efficiency of more specialized teeth. Specialization in teeth is like specialization in tools. You can, for example, tighten almost any nut or bolt with an adjustable wrench, but it takes longer because the wrench is clumsier and has to be fitted to the particular size of nut or bolt. You can tighten a nut much more quickly with an open-end wrench of a particular size, but only if it is of the right size. Our own back teeth are of this bunodont type (think of all the different things you eat); so are the teeth of pigs, those infamously omnivorous animals (if you think of it, however, people eat many things that any self-respecting hog would refuse!).

To be really effective for plant eating, however, the bunodont tooth must be further adapted. Considerable adaptation is evident in the teeth of grass-eating animals. The teeth must acquire a nearly flat grinding surface capable of taking considerable wear. Yet the surface must be kept rough instead of taking on a completely smooth finish that would be an inefficient grinder. The problem of maintaining a rough surface is solved already by the several materials that go into a mammal tooth. The outer covering is made of enamel—extremely hard but brittle. It will stand up as ridges under wear if it can be kept from chipping. Inside the enamel is the tougher but softer dentine, which wears faster than the enamel. When dentine wears, it forms low places in the teeth. Unfortunately, the dentine is inside the enamel and can support it only on one side. How do you protect the enamel from chipping? The answer lies in still a third material—the cement. In primitive mammals, and in human beings, this cement is confined to the root. This material welds the tooth to the socket (hence the name) and makes extraction so painful. In advanced plant-eaters, however, the cement erupts from the root region to cover most of the crown until wear exposes the underlying dentine and enamel. The cement is similar to dentine, tough but soft and keeps the enamel from splintering off from the outside. Thus, on the teeth of a plant-eater, at least if it is a highly developed form, you would expect to find three materials exposed on the grinding surface. The enamel, light-colored and shiny, stands up as ridges outlining the cusps. Inside the cusps is the softer, darker, but still shiny dentine. Filling in between the cusps is the dull and soft cement.

Another factor goes into the natural design of a herbivore's molar

tooth. The efficiency of the grinding surface is increased as the rough-
ness of the surface is increased. This efficiency demand required the
increase in length of the enamel ridges, a problem only partly solved
by the complex folding of the enamel found in some herbivores (like
the horses). Two different expedients have been developed by plant-
eating mammals. One is to connect adjacent cusps by ridges, giving
rise to a pattern termed *lophodont* ("ridge tooth"), found in the horses
and their relatives, and in the elephants. The other expedient was to
expand each cusp into a ridge in itself, usually crescent-shaped. Such
teeth are termed *selenodont* ("moon tooth") from the half-moon shape
of the worn cusps. Selenodont teeth are found in the cloven-hoofed
animals, like deer, camels, and cattle.

The rapid spread of the grasses, which seems to have occurred during the Miocene Epoch, opened up a new world to the plant-eating
mammals, a world of abundant food on open plains where predators
could be seen early and usually avoided. This world required the
solution of a serious problem before its advantages could be fully
exploited. Grass is very gritty; a grass-eating mammal is subject to
terrific tooth wear, wear so intense that a normal mammal tooth would
be long since worn down to the gums before the animal reached
maturity. This wear is heightened by another peculiarity of grass. It is
extremely tenacious of its nutritive value; it must be very thoroughly
ground before it yields up its food value.

The main solution to this problem lay in an extreme modification of
the teeth, making them extremely tall-crowned to allow for considerable wear. Because such tall teeth could not be exposed completely
above the gums, a large part of the crown must lie buried and be lifted
into place as wear occurred. The grass-eating rodents (and the rabbits,
which are lagomorphs, not rodents) found the ultimate solution; their
cheek teeth grow perpetually in many forms. Most of the other herbivores, especially the large ones, developed an expedient of growing
the tooth roots slowly, so that the crown is slowly lifted upward until it
has worn out. By this time, the animal has lived a long and productive
life and is well past breeding age. Both lophodont (such as horses) and
selenodont (such as cattle) teeth proved amenable to this solution.
The elephants found still another toothy solution to the problem,
which will be described a bit later.

The cloven-hoofed animals mostly have gone beyond this dental
solution to the problem of grass-eating and have developed a further
adaptation that eluded their rivals, the horses and the elephants. This
new idea may well be the key to the present dominant position of the
cloven-hoofed animals among the world's herbivores. The stomach
became divided into compartments. The first of these compartments
was a receptacle for half-chewed grass, so that the animal could cram

down a large quantity in a short time, before anyone else got to it. This receptacle (called the *rumen*, hence *ruminant*) was not just a holding chamber. There digestive enzymes and bacteria go to work on the grass, partly breaking it down and softening it. When stuffed, the animal could retire to a comfortable and safe place and lie down in the shade. The grass contained in the rumen could then be regurgitated a mouthful at a time and chewed more thoroughly at leisure and in its softened condition. Only when it was completely treated in this fashion would the food be passed on to the rest of the digestive tract for final processing. Thus was the utmost gotten out of every mouthful. An idea of the efficiency of this system can be gained from an examination of horse and cow manure—a distasteful but highly informative undertaking. Horse manure consists very largely of quite recognizable—and therefore undigested—bits of grass. Cow manure is much more fully broken down.

As an aside, this inefficient digestive system may partly explain why horses have never been raised on a large scale for food animals. Cattle are simply more efficient users of the available pasture. Social taboos resulting from the position of horses in human culture and the comparatively greater difficulty of controlling a given number of horses over the same number of cattle complete the picture. Many food taboos may have their real roots in ecological efficiency. The pork prohibition of some cultures may stem not from trichinosis prevention but from the fact that a hog eats as much of the same foods as does a human. It is then more ecologically sensible to feed the human, not the hog.

Obviously, plant eating is not the only specialization among the mammals. They have become adapted for exploiting almost every source of food on earth, and their teeth reflect these adaptations. The toothed whales (except for the most primitive) have developed simple fish-catching teeth. Their relatives, the whalebone whales, have dispensed with teeth entirely (except in the embryo) and use strainers grown from the ridges on the roof of the mouth to filter out tiny marine animals. The carnivores have developed their teeth into very specialized tools. The front teeth (incisors) retain their primitive nibbling role. The next teeth, the canines, also retain and enhance their early role as stabbers. The cheek teeth partly become powerful shears for cutting meat and tendons, a character particularly developed in the cats. Watch a dog or cat tackle an especially tough piece of meat; it will use the side of the jaw, not the front teeth. A particular tendency among the carnivores is the shortening of the jaw, carried to an extreme in the cats. This adaptation moves the shearing teeth (called *carniassals*) closer to the back of the jaw where leverage develops the most power. The cats are almost exclusively meat-eaters and have dispensed with grinding surfaces almost entirely; other carnivores retain, and may even further develop, grinding capabilities to allow a more varied diet. A few mammals, like the anteaters, have come to depend on soft-bodied insects for much of their food, things like termites and grubs. Their teeth tend to be reduced to simple pegs, often without enamel.

The most extreme dental adaptations are found in the gnawing mammals, the rodents and the rabbits. The incisors become ever-growing chisels. Because enamel with its slow wear is confined to the outside of the incisors and dentine to the inside, these teeth are self-sharpening. The incisors will grow whether they are used or not. A total denial of opportunities to gnaw is invariably fatal to a rodent, as the growing incisors will prop the mouth open and prevent any eating. Even more tragic is the loss of a lower incisor. The matching upper may even grow into a semicircle, piercing the roof of the mouth and perhaps even growing into the brain. Death will come eventually, from brain injury, infection, or starvation. As noted above, the cheek teeth may also be ever growing.

The mammals arose during the Triassic Period, from the synapsid reptiles. At various stages, the synapsids attained almost all the characters that we associate with mammals: legs under the body, complex teeth. Some may have been warm-blooded, may have had hair, and may have nursed the young. We cannot discern such characteristics from fossils. Until recently, the skeletal distinction between mammals and reptiles was based on the structure of the lower jaw. The lower jaw of reptiles is made up of several bones. In particular, the joint between the jaw and the skull is formed by the articular bone of the jaw and the quadrate bone of the skull. Sound is transmitted from the eardrum to the middle ear by a single bone, the stapes ("stirrup"). Mammals have only a single bone comprising the lower jaw, the dentary, which is jointed to the squamosal bone of the skull. The articular and the quadrate are not completely lost, however. They are pressed into service in the middle ear, becoming the malleus and the incus ("hammer" and "anvil"). The series of three bones, malleus, incus, and stapes, forms a set of magnifying levers for the transmission of sound from the eardrum to the inner ear. This greater sensitivity has allowed the eardrum to be buried in the skull at the end of a long tube for efficient protection of this necessarily delicate and vulnerable mechanism.

This simple distinction has now broken down, because of the tremendous amount of work that has been devoted recently to the fascinating story of the transformation of a reptile into a mammal. Considerable detail is now known of the skull anatomy of *Eozostrodon* (*Morganucodon* of many workers), a tiny creature from the Late

Triassic of Britain and China. This form straddles the reptile-mammal boundary nicely. The dentary bone has grown upward and backward to meet the squamosal on the outside of the jaw. Inside, the tiny articular still meets the quadrate to form part of the jaw joint. This arrangement is an exact intermediate between the reptilian and the mammalian condition. We now draw the line between mammals and reptiles at the point when the dentary meets the squamosal, thus making *Eozostrodon* "officially" a mammal.

Paleontologists have a peculiar ambivalence in their attitude toward animals like *Eozostrodon*. This ambivalence comes from the fact that paleontology has, in part, two conflicting goals. We attempt to provide neat pigeonholes for every animal, mainly so we can talk about each animal in a neat, precise fashion. Yet we are also studying the continuous flow of change in animal life through time, a flow that we can only divide arbitrarily. As long as there are gaps in the flow as we know it (what the layman would call "missing links"), they make beautiful places to put our divisions, our partitions between pigeonholes. Then along comes a beastie like *Eozostrodon,* which lands quite inconveniently right in the middle of our partition, the precise position of which has been determined by much argument.

Our way of thinking does not allow us to do without the partition. We were quite happy to find *Eozostrodon;* its existence demonstrates that our previous ideas probably were quite correct. But subconsciously we resent the animal. We must now tear out the partition we so carefully erected. We can't do without it, for it has been too useful. But we have opened up a new "can of worms." On which side of *Eozostrodon* do we put the partition? Is it a reptile or a mammal? We can't place it according to our old ideas, unless we define it as a reptile by a very strict application of the old dividing line. Yet this line would obscure its position as the earliest thing we would really feel comfortable about calling a mammal.

Whenever a new find like *Eozostrodon* comes along, strange things begin to happen to classifications. The partition has been pulled out; various workers have different ideas about just where to replace it. Eventually, the partition usually gravitates toward the largest gap in our existing knowledge. That is the easiest place to put it, until something comes along to fill the gap. Such gaps in our knowledge of the transition from reptile to mammal hardly exist any more, so we arbitrarily pick a point to put it, squeeze the adjacent forms aside a little, and drive it in. Essentially this way of categorizing is what has been done with *Eozostrodon.*

This discussion brings us to that ever present concept of the "missing link." Paleontologists don't like that term very much; we know

better. To show you why, let's indicate a line of descent by a row of
dots:

. . . .

The row illustrates the way it really is, but at this time we know about
only a few of the dots:

.   ...

Notice the big hole in the middle, which would be a "natural" place to
put a dividing line (/) if we had to "classify" this line of dots.

.   ...   /   ....

Now a new find comes along that belongs somewhere in the middle of
the gap, something that would be hailed in the popular press as a
"missing link."

.   ...

Our dividing line was not pinned down to any particular place; it just
sat somewhere in the gap. The first problem to solve is on which side
of the new dot our dividing line belongs. Or do we put this "missing
link" in its own category "/ . /"? Eventually this problem will be
ironed out to everybody's satisfaction (well, almost everybody's; there
always will be some mavericks).

The important thing to recognize is that our much-heralded "missing link" has not really solved any problems; in fact, we now would be
interested in exactly what lay on either side of it. Instead of one "missing link," we now have two. Here is why we really don't like to call
things "missing links." Remember: *every time you find a missing
link, all you have really done is to increase the known number of
missing links by one.*

Let's now go back to our original line of dots, which would represent a reasonably well-known lineage:

. . . .

Now, where do you put a dividing line between two groups that used
to be quite distinctive? You cram it in somewhere, between dots, and
it does not look "natural" anymore.

. . . . / . . . .

The problem becomes even worse when you realize that our row of
dots isn't a true picture. If we knew of all the animals that lay in this
line of descent, this is how our dividing line would look:

———————————————————————— / ————————————

Add the side branches that probably would exist on any such line of descent (mentally, please), and you will begin to recognize the corner into which classification has backed itself! And the problem will become worse the more we really know.

Yet back in the introduction to systematic paleontology, when we outlined the Linnaean system of classification, we came down pretty solidly on the side of traditional methods. Why? For one reason and one reason only: *until now it has served, and is still serving, a useful purpose.* The troubles of a classification system are more philosophical than real—we just don't know enough about complete lineages to place this system in real trouble—*yet.* When the time comes that the Linnaean system has outlived its usefulness, it should and will be abandoned. A few workers now think that time is already past. We think it is far in the future, but that future is coming closer every second and cannot be stopped. The time is upon us to begin seeking viable alternatives, while we still have time to make an orderly transition.

The fossil record of the mammals in Alabama is very scant. We have only vague indications of the rich and varied mammal life of Alabama's past. There are vast gaps; for example, not a single fossil rodent is known from the state. The techniques needed to collect small forms simply have not been applied on a large scale. Many other orders of mammals likewise are unknown in Alabama, yet they must have been present. The only order that can be considered really well known through organized collecting in Alabama is the Cetacea (whales).

Order Carnivora
"meat-eaters"

We put the term "meat-eaters" for the carnivores in quotes, as not all meat-eating mammals belong here (including ourselves), and not all members of the order are strictly meat-eating. For example, the bears largely have lost the habits involved in a completely carnivorous diet. Still, the name is appropriate. Most members of this order are at least somewhat specialized for a meat diet, with the speed and weaponry required, teeth for slicing meat and even bones, and rather simple digestive systems adequate for highly nutritious and easily digested meat. Also, most large, and even medium-sized and small, mammals with this type of diet belong in this order.

The hallmark of the true carnivore is in its carniassal teeth. These lie in the cheek region and are designed to cut meat and sinew. In true carnivores, these shears are always the fourth premolar (P^4) above and the first molar (M_1) below. Some earlier attempts by the mammals to become carnivores are not closely related to the true carnivores. These all had their carniassals farther back in the jaw, where they developed more power, but were harder to reach. These are usually split off as the Order Creodonta. None of these as yet have been found in Alabama, though the whales probably are descended from some of them.

Another feature of carnivores is the presence of large canines (eye teeth). These are used in catching and killing prey, aided by the sharp claws. In some, especially the saber-toothed cats, these canines became huge, slashing blades.

FAMILY MIACIDAE

The miacids are a very primitive group of extinct small carnivores and are almost certainly ancestral to living forms. The only known land mammal, first reported here, from the Eocene of Alabama, belongs here. The specimen is only half a tooth, a lower carniassal, from the Gosport Formation at the well-known Little Stave Creek locality in Clarke County. It is a tribute to the distinctive teeth of mammals that it can be identified as *Vulpavus* sp. (identification credited to Bob H. Slaughter of Southern Methodist University). As half a tooth normally would mean little to most people, we figure here (fig. 77) a complete cheek dentition from the West. This find has whetted our appetites for more—and they are surely there.

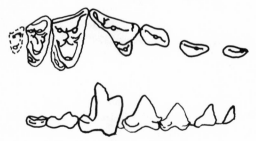

FIG. 77. *Vulpavus* sp., upper and lower right cheek teeth. Uppers in occlusal view, lowers in lateral view. x2. This is a western specimen. The Alabama specimen is only about half the size of the largest lower tooth. After Matthew.

FAMILY URSIDAE
bears

A fragment of bear jaw was recovered from a cave in Lauderdale County, Alabama, during a survey of Alabama caves undertaken by the Geological Survey of Alabama, The United States Geological Survey, and the National Speleological Society. The fragment may or may not be old enough to be considered a fossil.

Order Edentata
armadillos, sloths, anteaters

The living edentates are really little but stragglers of a vast and varied group that developed in total isolation in South America during most of the Tertiary. Many other strange mammals belonging to other orders also evolved in this refuge, including large carnivorous marsupials (one of them closely resembled the saber-toothed tiger in its adaptations) and a vast assemblage of peculiar hoofed animals. When North and South America were joined in the late Pliocene by the emergence of the Isthmus of Panama, many North American forms invaded the southern continent, and most of the highly specialized South American fauna rapidly vanished under the pressure of competition. Only the edentates survived and flourished through this change and even managed to stage a successful counterinvasion of North America.

The anteaters and tree sloths do not seem to have participated in this invasion-in-reverse. At least, their fossil remains are not known in North America. However, they seem always to have been forest dwellers, and the fossil record of this habitat is notoriously scanty. The armadillos came, were driven back at the close of the Pleistocene, and are now staging a remarkably successful second advance. Other invaders that did not survive the close of the Pleistocene were the glyptodonts, giant armadillolike forms that had a solid back armor and reached lengths of over 10 feet, and the giant ground sloths (see below).

The whole edentate skeleton is exceedingly peculiar (see fig. 78). We have not the space to describe it in detail, but some outstanding points can be described quickly. The vertebrae have strange extra articulations between them. The scapula (shoulder blade) has a loop of bone that joins the acromion (the ridge down the outside of the scapula) to the most ventral part of the front edge, found in no other mammal. The teeth are invariably simple. The early edentates seem to have been eaters of soft-bodied insects and larvae, and the cheek teeth became simple pegs, often without a covering of enamel. This form persists even in those edentates that later became plant eaters, though the sides of the teeth may show some infolding of the enamel (in the glyptodonts) and the teeth become high-crowned, perhaps ever growing.

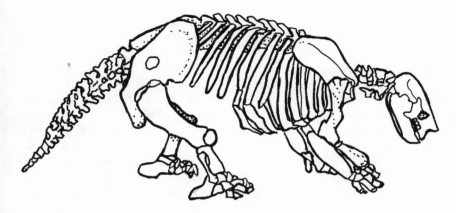

FIG. 78. *Mylodon*, reconstructed skeleton, length of original, about 3 meters. Almost every bone is highly distinctive. After Gregory.

Suborder Pilosa
hairy edentates—sloths and anteaters

This group seems to be sharply separated from their armored relatives by the presence of thick, shaggy hair instead of solid armor plating, perhaps with hair between the plates. Some representatives do have nodules of bone imbedded in the hide but never to the extent of joining into a complete armor. All of them have a peculiar gait, as the front feet have enormous claws that would interfere with normal walking. The ground-dwellers walk on the "knuckles" of the forefeet, while the tree sloths bypass the problem by spending virtually all their time hanging upside-down from branches by their hooklike claws. The tree sloths are almost helpless on the ground.

FAMILY MEGALONYCHIDAE
giant ground sloths, in part

There are three families that can be lumped under the name "giant ground sloth." These are the Mylodontidae ("millstone-toothed"), the Megatheriidae ("giant beasts"), and the Megalonychidae ("giant claws"). None of these names are particularly appropriate as each could be applied to the group as a whole. The families are distinguished on details of the teeth and skeleton. All three invaded North America during the late Pliocene and the Pleistocene, and all became extinct at some time near the end of the latter, at least in North

America. The largest were larger than a modern elephant, reaching a length of 20 feet (Romer, 1945, p. 484) and, standing on its hind legs, looked out upon the world from a height of about 15 feet.

Considerable evidence indicates that ground sloths held out in isolated pockets until quite late. A rather small species in the West Indies may even have been exterminated by the Spaniards shortly after their arrival. There are persistent rumors of a ground sloth skin, complete with its scattered bony plates, hung to dry on a barbed-wire fence somewhere in Argentina, but these rumors are unsubstantiated. Until someone goes out and bothers to pick it up, or better yet, finds a fresh specimen, rumors of living ground sloths belong in the same category as those of Loch Ness monsters and abominable snowmen. Science, at least paleontology, deals in *specimens*, not rumors or speculations.

Megalonyx jeffersoni (Desmarest) (fig. 79)

Megatherium jeffersoni Desmarest, Mammalogie, p. 366 (*fide* Hay, 1902A, p. 578).
Megalonyx jeffersoni, Leidy, 1855B, p. 6, pl. 16, fig. 13; Tuomey, 1858A, p. 15; Mercer, 1897, p. 38; Hay, 1923, p. 40.

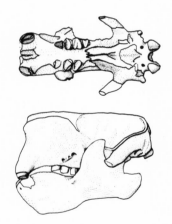

FIG. 79.   *Megalonyx jeffersoni* (Desmarest). Skull, lateral and ventral views. Locality not given (?Kentucky). After Müller, 1970. xc. ⅛.

The list of references under the name *Megalonyx jeffersoni* looks impressive in numbers, but all refer to the same bones. These came from a cave near Tuscumbia, Alabama, (where caves are numerous) and were collected by a Dr. Powel. From him, the fossils passed to the first State Geologist of Alabama, Michael Tuomey, and thence to Joseph Leidy and the Academy of Natural Sciences of Philadelphia. Leidy (1855B) gives the locality incorrectly as "Tuscumbia County,

Alabama," but this was corrected to Colbert County by Tuomey (1858A).

The jaws are very deep and massive. There are four lower and five upper teeth, and the front tooth in the upper jaw is caniniform, that is, it resembles the canine tooth of other mammals. The cheek teeth are roughly rectangular, very tall prisms with fluted sides. The enamel covers the sides, but the interior is a solid plug of massive dentine.

This species is named for President Thomas Jefferson, who took an early and intense interest in fossil vertebrates.

Order Proboscidea
elephants and mastodons

The living elephants, like the living representatives of so many orders, are the last remnants of a vast and varied assemblage that begins in the Early Tertiary and flourished until a very few thousand years ago. The living elephants hardly need further description.

A peculiar feature of the elephants is their solution to the problem of tooth wear. The vast size of the elephants could only be attained through a very long life (maturity at about twenty years; perhaps living to 100 in rare cases); high-crowned teeth developed, but such teeth were not enough. Each tooth was adapted into a large, very complex structure with a vast grinding surface that accommodated food needs at that stage of life. Only one tooth is fully in place at any one time in each jaw half, but it is perpetually being crowded forward by a new tooth from behind. In succession there appear three milk teeth and three "permanent" molars. Each moves forward, is worn away to its base, and is shed piece by piece. Meanwhile, the next tooth is growing in, and its front part is in use while the back part of the preceding tooth is still functional. This succession gives an extremely long-lasting surface even when subject to much wear. The final development of this system is attained only in the Elephantidae.

FAMILY MASTODONTIDAE
mastodons

These extinct relatives of the elephant attained large size, at least that of a living Indian elephant. The teeth are quite distinctive and are replaced in the fashion outlined above, although more than one tooth is fully functional at once. Mastodon teeth attain a width of 4 to 5 inches (10 to 12 centimeters) and a length of about a foot (30 centimeters). Each tooth bears two rows of three to five powerful cusps, one lingual and the other labial. Each pair of cusps is connected by a powerful ridge (loph) running across the tooth. The enamel is exposed over the entire surface, except where worn away to show the dentine. Cement is confined to the roots and is not found on the crown.

Mammut americanum (Kerr) (fig. 80)

Elephas americanus Kerr, 1792, Anim. Kingdom, p. 116 (*fide* Hay, 1902A, p. 708).
Mastodon, Tuomey, 1858A, p. 15.
Mamut americanus, Hay, 1923, p. 124.

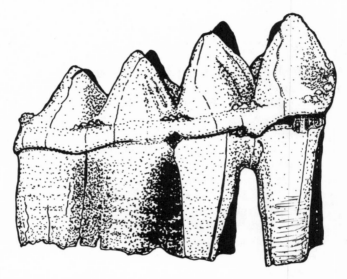

FIG. 80.   *Mammut americanum* (Kerr). Right 3rd mandibular molar. From Pleistocene deposit along Tombigbee River in Sumter County, Alabama, near Demopolis. University of Alabama Museum of Natural History. x½.

The teeth, the only common and identifiable fossils of this species, are described above. Tuomey (1858A) reports this form (as *Mastodon*) from a cave near Tuscumbia, Alabama, and indicates that the material comes from a different cave from that yielding remains of *Megalonyx* (see above). Hay (1923) reports fragments of teeth from near Bogue Chitto, Dallas County. A quick search of the collections of the Geological Survey of Alabama reveals specimens from Wilcox, Marengo, and Sumter counties. One of the authors (Jones) collected a skull and several ribs of this species along the Tombigbee River in Sumter County just west of Demopolis. This evidence should be sufficient to indicate that the species is widespread in the Pleistocene of Alabama.

Even fragments of enamel can be quite distinctive and easily identified as *Mammut*. The enamel is very thick (⅛ inch or more) and will show one smooth, highly polished side and a rougher, unpolished side. The smooth side is the outer surface. Enamel of elephantids will be rough on both sides.

One of the most peculiar results of taxonomic strictness is the use of *Mammut* as the proper name for this genus, as the name sounds more like it should belong to a mammoth (a popular name applied almost indiscriminately to any fossil true elephant). "Mastodon" seems indelible as a popular name, even among professionals.

FAMILY ELEPHANTIDAE
elephants and mammoths

The structure of an elephant tooth can perhaps best be described by giving a "recipe" for converting a mastodon tooth into one. First make a mastodon tooth, using gelatin for dentine and rubber for enamel. Don't give it four or five pairs of cusps, though; make it ten or more, perhaps even forty. And make the ridges between cusps almost as tall as the cusps. Now take the whole affair and put it endwise in a vise. Screw down the vise jaws until the cusp pairs are crushed almost into flat plates. Pack cement into all available openings. Let everything harden, then file down the top to a flat surface. What would you come up with? On the side, you would see long plates of enamel with cement between. On top, where it was filed off, the plates of enamel would become loops, much flattened, with dentine in the center and cement between adjacent loops. If you put short roots on it, and filed down the front end more than the back, you would have a fair approximation of an elephant tooth.

The elephants seem to be a late development from the mastodons, with primitive types in the Pliocene and fully developed forms by middle Pleistocene. They have thus been around for about the same time as man. A "mammoth" is nothing but a fossil elephant. Some attained enormous sizes, even by modern elephant standards. *Elephas ganesa* (named for the Hindu elephant-headed god Ganesa) from the Pleistocene is reported to have stood 18 feet at the shoulder. On the other hand, there are dwarfed island races, especially well known from the Mediterranean. Some adults of *E. falconeri* from the Pleistocene of Sicily stood little over a foot at the shoulder, with 6-inch tusks. Think how they would sell as pets!

Elephas columbi Falconer

Elephas columbi Falconer, 1857B, in table facing p. 319.

One of the original specimens of Columbian mammoth (but not the type specimen) came from an unspecified locality in south Alabama. This was almost certainly the specimen noted by Warren (1855, *fide* Hay, 1923, p. 165) from "Alabama, near Gulf of Mexico." Hay (1923, p. 165) also notes this species from Bogue Chitto, Dallas County but refers both to *Elephas imperator* Leidy (1858B, p. 10; Imperial mammoth).

The distinction between *E. columbi* and *E. imperator* is probably not to be made solely on isolated teeth. Hay (1923) used an old "rule of thumb" to distinguish these species. Measure off a distance of 10 centimeters (4.1 inches) on the tooth from front to back and count the number of loops of enamel in that distance. If there are 5 or less, put it in *E. imperator;* 6 or more, *E. columbi;* in between 5 and 6, call it *Elephas* sp. This method should be used only on last molars, teeth that are difficult to identify. Hay's rule of thumb is approximate at best, but it is better than nothing. We will follow a more conservative approach and call them all *E. columbi* until better material is available.

As noted above, elephant enamel is rough on both sides. There are numerous Alabama specimens in the collections of the Geological Survey of Alabama and the University of Alabama Museum of Natural History; they come from almost every part of the state, wherever river deposits of Pleistocene age are available.

Order Sirenia
manatees and sea cows

Siler (1964) reports a fragment of rib, probably sirenian, from the Gosport Sand (middle Eocene) in Monroe County, Alabama. Other specimens, also rib fragments, are under study by Clayton Ray of the United States National Museum. The bones of sirenians, especially the ribs, are distinctive in being made up of unusually compact bone, with very few and small openings internally. The reason for this massiveness of the bone is uncertain; it occurs in several groups of aquatic vertebrates and may serve as ballast. The fragment cannot be identified further.

The sirenians seem to be somewhat related to the elephants. At least, the teeth move and are replaced from behind.

Order Perissodactyla
the odd-toed hoofed animals—
horses, tapirs, rhinoceroses

The hoofed animals, or ungulates, are a far more complex group than appears at first glance and are now divided into several orders, largely distinguished by technical details. Only two of these orders survive in anything like a flourishing condition today, and two more, Proboscidea and Hyracoidea (the small, Old World cony or hyrax), survive in very restricted fashion.

One of the comparatively flourishing orders is the Perissodactyla, but even this has yielded much ground before the more successful Artiodactyla, or cloven-hoofed animals. The two orders can be distinguished by the structure of the legs and teeth.

The leg of a perissodactyl is built around an axis passing through the middle (third) toe. The main weight of the body rests on this toe, and the other digits are reduced or even lost. The horses represent the ultimate in this development. Only one toe in each foot is functional; the toes to either side are retained only as comparatively useless vestiges ("splint bones"). The main bone of the ankle region, the astragalus, has a powerful pulleylike surface at its upper end where it meets the main lower leg bone (tibia) but is almost flat below. The teeth are lophodont. The surviving perissodactyls are the horses (including asses and zebras), the tapirs, and the rhinoceroses.

Suborder Hippomorpha

FAMILY EQUIDAE
horses

The horses have one of the most extensive and best-known fossil records of any vertebrate group. The family began as small, forest-dwelling, browsing forms in the Eocene and slowly increased in size until the Miocene. The earliest known ancestor of the modern horse, *Hyracotherium ("Eohippus"),* was quite small, about the size of a fox terrier; his late Oligocene and early Miocene descendants may have reached the size of a large dog.

With the rapid spread of the grasses in the Miocene, the horses were among the first to exploit this environment. The legs became long and slender, and only a single toe was left functional at the close of this epoch when pony size had been reached. The theme of horse evolution has been the increase in size and speed; the principal defense of the horse always has been an ability to stay the proverbial "one step ahead" of predators. By the middle of the Pleistocene, an essentially modern size and structure had been attained, and the horse fauna was large and varied.

Then came a strange event in equine history. The whole story until then had been played out essentially in North America; the rest of the world "made do" with whatever horses managed to get there. Often these were not the "latest American model." Suddenly, about 9,000 to 10,000 years ago, the entire American horse fauna, with several species, disappeared, wiped out by forces we do not entirely understand as yet. They held on in Asia, Europe, and Africa (there as zebras) until reintroduced into the Americas by the Spanish in the sixteenth century. They were an almost instant success, indicating that whatever had wiped them out no longer existed. They spread so very rapidly that almost all Indian tribes had them by 1750, and their impact on the Indian way of life was terrific. Whole tribes of settled farmers and small-scale hunters took to a nomadic existence on the Great Plains, following the bison or buffalo and raising quick-growing crops on the few occasions when they settled down long enough. It is hard to realize that the way of life that we almost automatically think of as "Indian" did not, indeed could not, begin until white men replaced the long-lost horse.

The cheek teeth of later horses (the only ones found yet in Alabama) are tall, fluted prisms. The grinding surface of the upper teeth is nearly square; that of the lowers is oblong. The premolars are completely "molarized" and hardly can be distinguished from the molars. Only the first and last of the cheek teeth depart from this shape; their ends are somewhat tapered. The enamel ridges on the grinding surfaces form extremely complex patterns; these patterns are the most commonly used means of identifying a fossil horse. Careful attention to fine detail is necessary in this type of study, as is an appreciation of the modifications that can occur because of different stages of wear.

In extreme old age, a horse tooth loses its prismatic shape as the crown is worn down. Deep, forked roots appear as the last portions of the crown are lifted into place. Any horse tooth that shows such roots is most likely nonfossil; only domestic horses are likely to reach such an advanced age. A quick test is to hold the roots in a flame until they are charred, then quickly smell them. If you detect an odor of burned meat, indicating that there are still proteins in the bone, it is almost certainly less than a few thousand years old (with horses, certainly less than a very few hundred). This "match test" works well with other fossil bones.

Hipparion phosphorum Simpson (fig. 81)

Hipparion phosphorum Simpson, 1930H, p. 189, fig. 20A.
Hipparion phosphorum, Whitmore *in* Isphording and Lamb, 1971, p. 778.

This is a fairly large species for a Pliocene horse, though small by Pleistocene or Recent standards. The teeth are about an inch across, and the enamel pattern is shown in fig. 81. The species was described

FIG. 81.   *Hipparion phosphorum* Simpson. Illustration of the type (F.S. V-1423) after
Simpson, 1930H. Natural size.

originally from the phosphate beds of Florida. The Alabama occurrence is in the Citronelle Formation in northern Mobile County and
was reported by Whitmore (*in* Isphording and Lamb, 1971). The Alabama material has not been described, so we figure the type specimen
from Florida.

Nannippus sp. cf. *N. lenticularis* (Cope)

?*Protohippus lenticularis* Cope, 1893A, p. 41, pl. 12, figs. 1–2.
Hipparion lenticulare, Osborn, 1918A, pp. 26, 174, 184, pl. 32, fig. 2, pl. 33, figs. 5–7,
text-figs. 147, 148A.
Nannippus sp. cf. *N. lenticularis*, Whitmore *in* Isphording and Lamb, 1971, p. 778.

The teeth of this species are about the size of *Hipparion phosphorum* but have a somewhat simpler pattern. Its Alabama occurrence
is the same as that of *H. phosphorum*.

Equus sp. (fig. 82)

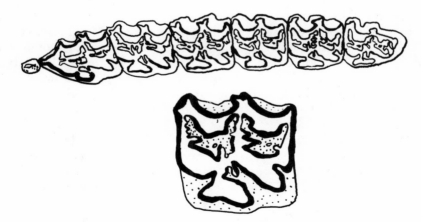

FIG. 82.   *Equus* sp. (modern horse). Upper left cheek teeth in occlusal view, x½ (approximate), showing enamel pattern. Below: first upper left molar, x1 for
medium-sized horse. Dark lines are enamel, white areas dentine, stippled
areas cement. After Gregory.

Hay (1923) has two reports of isolated horse teeth from the Pleistocene of Alabama. One is an upper right molar from Bogue Chitto, Dallas County, collected by L. C. Johnson in 1883. Hay (1923) identifies this specimen as *Equus leidyi* Hay (1913B), but current usage tends to shy away from identifying a Pleistocene horse on a single tooth unless it is quite distinctive. One now waits for more complete dentitions and hopes for several specimens before attempting precise identification. There are numerous specimens, mostly with scant locality data, in the collections of the Geological Survey of Alabama and the University of Alabama Museum of Natural History. Some upper teeth (locality unknown) in the latter collection may be large enough (about 1½ inches across) to refer to the largest Pleistocene horse from North America, *E. scotti* Gidley (1901A).

Suborder Tapiromorpha
tapirs and rhinoceroses

These are generally heavy-bodied perissodactyls. They never reduce the feet to a single-toed condition; there are three or four functional toes on the front foot and three on the hind. The teeth are not particularly high crowned and almost never show cement on the surface; they may, however, be extremely massive.

FAMILY RHINOCEROTIDAE
rhinoceroses

The most obvious feature of a living rhinoceros, other than its massive size and short temper, is the nose horn, whether single or double. Don't expect to find it in a fossil. Rhinoceros horn is really composed of solidly fused hair and lacks a bony core, unlike a cow's horn. It is not amenable to fossilization except under very special circumstances. The best you can expect to find is a roughened and reinforced area on the skull for its attachment; besides, many early rhinoceroses lacked horns.

The teeth are large and massive with a fairly simple pattern of ridges. On the upper teeth, this pattern is more or less in the shape of a Greek letter π; on the lowers, the ridges resemble two very distorted crescents, one of which joins its end to the other's center.

Teleoceras sp. (fig. 83)

This is a typical North American rhinoceros of the Miocene and early Pliocene. It is a long-bodied, short-legged, low-slung beast that

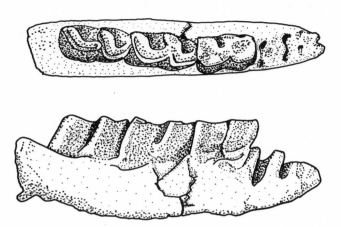

FIG. 83. *Teleoceras proterus* Leidy. Right lower jaw with M 1-3, crown and external
views. After Simpson, 1930H, fig. 22. x⅓.

can be referred to (tongue-in-cheek) as a "sports-model rhino." The
horn was probably small, as its attachment is not large and prominent.

Teleoceras was reported along with other lower Pliocene vertebrates from the Citronelle Formation of northern Mobile County by
Whitmore (*in* Isphording and Lamb, 1971).

Order Artiodactyla
cloven-hoofed animals

These are the most successful of the medium-to-large plant-eaters
today. One of their branches, the ruminants, is particularly prominent
for reasons discussed above (in the introduction to mammals).

The teeth are bunodont (with rounded cusps) in primitive forms and
selenodont (crescent-cusped) in more advanced types. The astragalus,
the main ankle bone, is pulley-shaped at both ends instead of just one,
a unique feature among the mammals. Instead of passing through the
third toe, the main load-bearing axis of the foot goes between the third
and fourth toes. Thus the foot never develops a single finger or toe;
there are normally two main load-bearing toes leading to the "cloven
hoof" so characteristic of the order. The toes to each side retain hooves
but are shortened; they are of use mainly on soft ground. However, the
metapodials (long bones of the hand and foot) of the main toes often
are fused into a single bone (cannon bone) that bears a pair of "rockers" (joint surfaces for the toes) at its lower end and retains traces of
its dual origin by having a double marrow cavity.

The bunodont artiodactyls comprise the pigs and their relatives—
the families Suidae (swine), Hippopotamidae, and Tayassuidae (pec-
caries or javelinas). Only the last of these is known in the fossil record
of North America (the other two were exclusively Old World until
people began shipping hogs to every part of the world), but it has not
been reported yet in Alabama. These three families, plus some extinct
relatives, comprise the Suborder Suina. Some extremely primitive
Eocene forms are split off into a suborder Palaeodonta.

Suborder Ruminantia
ruminants

All representatives of this suborder "chew the cud" in the Biblical
sense. These include any animal that "parteth the hoof, and is cloven-
footed, and cheweth the cud" (Lev. 11:3), and all would be fair game
for the hungry but pious Hebrew. For some reason, camels were
excluded (Lev. 11:4), apparently because our Biblical taxonomist re-
fused to recognize that they part the hoof, perhaps because the foot is a
spreading pad in which the hooves are comparatively unimportant.
Even so, we must include camels in this suborder.

All the ruminants have selenodont teeth that may be either high or
low crowned. There is always a cannon bone in the leg.

FAMILY CAMELIDAE

Camels are another group recently added to our store of knowledge
about Alabama bones by Whitmore (*in* Isphording and Lamb, 1971).
We do not have detailed descriptions of the material; however, the
cannon bone is highly distinctive and commonly preserved. It is
somewhat splay-ended because the two fused bones separate some
distance above the rockers. A median ridge extends around the articular end of the rocker, as in all other ruminants, but it does not go
completely around from joint-edge to joint-edge. The material was
from the Pliocene Citronelle Formation of northern Mobile County. It
has not been identified further.

FAMILY PROTOCERATIDAE

This is a family of strange, deer-sized ruminants, known only in
North America from the Oligocene through the Pliocene. They bear a

weird and wonderful assortment of horns, numbering as many as six. Their number and position are quite varied from genus to genus.

Synthetoceras sp. cf. *S. tricornatus* Stirton (fig. 84)

?*Synthetoceras tricornatus* Stirton, 1932E, pp. 147–168, 3 figs., 5 pls.
Synthetoceras cf. *S. tricornatus*, Whitmore, *in* Isphording and Lamb, 1971, p. 778.

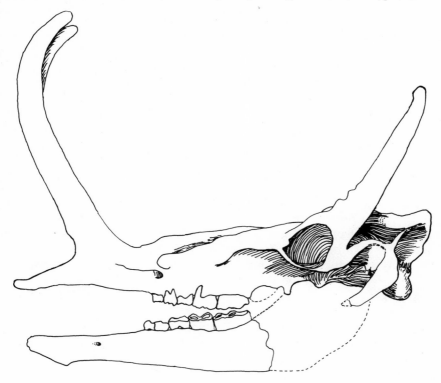

FIG. 84.   *Synthetoceras tricornatus* Stirton. Skull and mandibles. Frick AMNH 33407 (skull) and 32468 (mandible), Clarendon, Texas. Alabama specimens of this or a related species are fragmentary. After Frick, 1937. x¼.

Chalk up another one from the Citronelle Formation of northern Mobile County, Alabama! This is an unusual fellow, even for a protoceratid. The skull bears three horns; two are cowlike, curved, and situated by the ears; the third is quite long, even up to a couple of feet, and situated on the nose. As if this pattern were not weird enough, the nose horn is forked near its end.

FAMILY CERVIDAE
deer

We are now back on more familiar ground, though we are still depending on the Citronelle Formation of northern Mobile County (and, of course, Whitmore *in* Isphording and Lamb, 1971) for our Alabama fossil record.

Distinctive of deer are the antlers—complex, hornlike, often branching growths that are shed each fall and grown again in the spring. In most deer, the antlers are covered with skin and hair ("velvet"), which is rubbed off after the antlers are fully developed. During courtship competition males fight it out with antlers and hooves; after the mating season, the antlers are shed. Antlers are normally present only in the males (a fact known to all hunters), except among the Arctic reindeer and caribou, both sexes of which bear antlers.

The teeth are selenodont but very low-crowned, with the roots forking right at the gum line. Deer are browsers on soft forest vegetation rather than being mainly grazers. This eating habit is indicated by their low-crowned teeth.

A superficially similar family is the Antilocapridae, now surviving only in the pronghorn "antelope" of the Great Plains. However, the pronghorns have true horns consisting of bony cores with an outer covering of horn. Strangely enough, the horn covering is shed yearly, though the bony core remains, distinguishing true horn from antlers, which are shed completely each year. The teeth are high-crowned and go deep below the gum line before forking into roots. Some Pleistocene forms *(Tetrameryx* and *Stockoceros)* had four horns. No antilocaprids are known in the fossil record of Alabama, nor in the living fauna, but we would not be greatly shocked (though considerably pleased) if you came up with one.

FAMILY BOVIDAE
cattle and relatives

This is the most important group of living grazers. The teeth are very high-crowned but retain the selenodont patterns. The horns are usually massive and never shed. There are never more than two horns, except in some peculiar domestic breeds of sheep.

Included here, in addition to cattle, are bison (buffalo), musk oxen, sheep, goats, and the true Old World antelopes. They are a fairly recent group, the earliest forms appearing in the Miocene, a very appropriate time for a group of grazers.

Bison sp.

Until the summer of 1976, when a fine specimen of a partial skull was found in Montgomery County, only a single tooth of a fossil bison (buffalo) was known from Alabama. The skull of *Bison antiquus* (Leidy), found by Mr. Mendel Dantzler of Greenville, Alabama, measures 44 inches from the tip of one horn core to the tip of the other.

Teeth of bison are extremely difficult to distinguish from cow teeth, often even for experts. The minimum required to attempt identification is a nearly complete horn core. These may be over 3 feet long in some fossil species. Bison did not appear in North America until fairly late in the Pleistocene, probably about seventy-five thousand years ago. They were the only large grazers to survive the close of the Pleistocene in most of North America, thus accounting for their fantastic abundance on the Great Plains in historic times. They were a prime food of the earliest known North American hunters and later of the Plains Indians after they acquired the horse for riding.

Order Cetacea
whales

We have saved the biggest for last. Even the largest dinosaurs almost shrink into insignificance next to the vast bulk of the living blue whale, which perhaps attains a weight of 150 tons.

We should have put the above in the past tense. The day of the great whales is coming to an end because of human greed. Every year the number of whales diminishes, while whaling proceeds at the same rate, with ever more ruthless efficiency. Two species, the right whale and the bowhead, were almost wiped out early. Though hardly hunted for nearly a century, they are still a rare sight in the Atlantic where they once sported their portly dignity by the thousands. They may never recover. The sperm whale may just be holding its own, with luck; demand for its special products has diminished. The great blue whale and its slightly smaller relative, the finback, are now the most threatened. No longer can the psalmist sing: "So is this great and wide sea, wherein are things creeping innumerable, both small and great beasts. There go the ships: there is that leviathan, whom thou hast made to play therein [Ps. 104:25–26]."

We do not have the right to destroy these marvelous creatures. Before it is too late, we must take to heart the preceding verse: "O Lord, how manifold are thy works! in wisdom hast thou made them all: the earth is full of thy riches [Ps. 104:24]."

Suborder Archaeoceti
primitive whales

Much of the story of early whale development has been deciphered in Alabama. The upper Eocene Jackson Group is rich in their remains, especially in Clarke, Choctaw, and Washington counties and adjacent areas in Mississippi. We cannot do justice to the subject here; Kellogg (1936A) is an excellent and thorough treatment.

The archaeocetes are a very primitive group of whales. The skull is only slightly changed from the primitive mammalian pattern; there is no trace of the "telescoping" that occurs in the skull roof of the advanced whales. The teeth are extremely complex, with many cusps in a line along the edges of a great triangular blade. The body seems to have been remarkably slender compared to the stoutness of modern whales. The earliest known forms are from middle Eocene deposits in Egypt, and the only other good material is from Alabama. The latest known are scrappy specimens from the Miocene of Europe, but the suborder was not prominent after the close of the Eocene.

The literature on the archaeocetes of Alabama is vast and much of it exceedingly difficult to find. We will cite parts of it in our synonymies, but our treatment will be far from exhaustive.

FAMILY DORUDONTIDAE

These archaeocetes are stout-bodied and not so large as the basilosaurids (see below).

Zygorhiza kochii (Reichenbach) (figs. 85, 86)

Basilosaurus kochii Reichenbach, *in* Carus, 1847, p. 13, pl. 2, figs. 3–4.
Zeuglodon brachyspondylus minor Stromer, 1903D, p. 85.
Zygorhiza minor, True, 1908, p. 78.
Zygorhiza kochii, Kellogg, 1936A, pp. 101–173, figs. 29–75.

This is a small archaeocete from Alabama. The type specimen was part of the infamous Dr. Koch's *"Hydrarchos"* "skeleton" and was soon recognized as distinct when the material passed out of his hands. Kellogg (1936A) believes the type came from near Clarkesville, Clarke County, Alabama. The vertebral centra are about as long as tall, not elongated as in the basilosaurids. The vertebral processes seem to be about "normal" size for the vertebrae and are long enough to have articulated with adjacent vertebrae. The body was much stouter than that of the basilosaurids; *Zygorhiza* thus has considerably more potential as an ancestor of the later whales than do the larger and more spectacular forms.

Another species, closely related, is *Dorudon serratus* Gibbes (1845A), but Kellogg (1936A) does not believe that material of this form has been found in Alabama.

FAMILY BASILOSAURIDAE
zeuglodonts

The zeuglodonts are essentially a very long-bodied family of Eocene archaeocetes, with a few Oligocene stragglers. The skull, as in the other archaeocetes, shows an essentially primitive mammalian pattern; the nostrils are at the front of the skull. The teeth can be readily differentiated into incisors, canines, molars, and premolars. The cheek teeth bear multiple cusps.

The vertebrae, except in the neck region, are enormously elongated, about twice as long as high in the centrum. The neural arch did not keep pace with this vertebral growth, and is scarcely larger than that of *Zygorhiza*. As a result, the neural arches seem to perch in almost ridiculous fashion atop a centrum that is several sizes too large, like a small child on a draft horse.

Basilosaurus cetoides (Owen) (figs. 85, 86)

Basilosaurus Harlan, 1834C, pp. 397–403, pl. 20, figs. 1–3 (no species named).
Zeuglodon cetoides Owen, 1839B.
Zeuglodon cetoides, Owen, 1841C, p. 69, pls. 7–9.
Zeuglodon harlani De Kay, Zoology of New York, 1842, pt. I, Mammalia, p. 123 (*fide* Kellogg, 1936A, p. 15).
Hydrargos sillimani Koch, 1845A.
Zeuglodon ceti Wyman, 1845, pp. 65–68.
Hydrarchos harlani Koch, 1845B, pp. 1–24.
Zeuglodon macrospondylus Müller, 1849A, p. 21.
Alabamornis gigantea Abel, 1906, pp. 450–458, figs. 3–4.
Basilosaurus cetoides, Kellogg, 1936A, pp. 15–100, figs. 1–28; all the above citations, except the first and last, are *fide* Kellogg, 1936A, p. 15.

The longest, if not the largest, of the Alabama archaeocetes, this species is also the form that a vertebrate paleontologist almost invariably will think of first when "Alabama" is mentioned. A few seem to think that *Basilosaurus cetoides* is the only fossil vertebrate found in Alabama, but we hope this myth is dispelled by now. The beast attained a length of 70 feet, with a 5-foot skull, so that Koch's "reconstruction" (see Introduction) was not too much exaggerated. See above for a description.

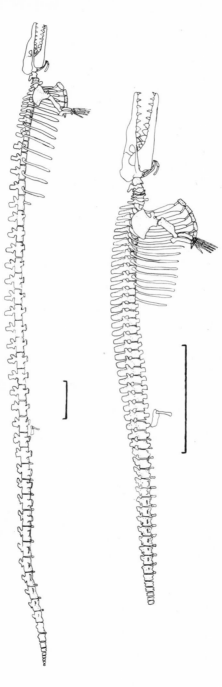

FIG. 85.   Reconstructed skeletons of Alabama archaeocetes. Top: *Basilosaurus cetoides* (Owen), based on three specimens from Choctaw County, Alabama. Scale bar 1 meter. Bottom: *Zygorhiza kochii* (Reichenbach), based on several specimens from Clarke and Choctaw counties, Alabama, and Clarke County, Mississippi. Scale bar 1 meter in decimeters. Skulls in outline only, see fig. 86. After Kellogg.

Because this species is so completely associated with Alabama, it is difficult to recognize that Harlan's (1834C) type specimen was from Caldwell Parish, Louisiana. The specimen, a single vertebra, is still preserved at the Academy of Natural Sciences of Philadelphia, number 12944A. The type specimen of Owen's species (the first named) is from Clarke County, Alabama, on land then belonging to Judge John G. Creagh.

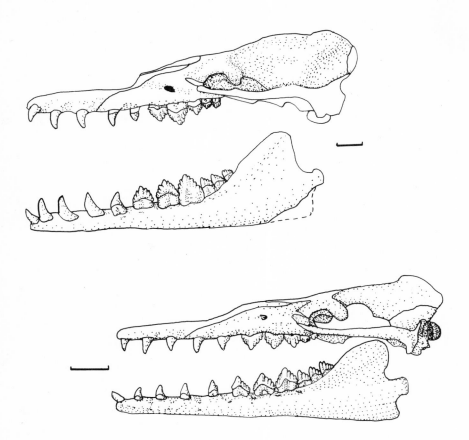

FIG. 86.   Skulls of Alabama archaeocetes. Top: *Basilosaurus cetoides* (Owen), based on USNM 4674, from near Melvin, Choctaw County, Alabama. This skull is used in the reconstructed skeleton at U.S. National Museum of Natural History, Washington. Bottom: *Zygorhiza kochii* (Reichenbach), based on USNM 11962, same locality. After Kellogg. Scale bars 10 cm.

Pontogeneus brachyspondylus (Müller) (fig. 87)

Hydrarchos harlani Koch, 1846, pp. 1–20 (in part).
Zeuglodon brachyspondylus Müller, 1849A, pp. 26–28, pl. 13, figs. 6–7.
Pontogeneus priscus Leidy, 1852, p. 52.
Pontogeneus brachyspondylus, Kellogg, 1936A, pp. 248–255, fig. 82j, pls. 36–37.

Koch's "reconstruction" yielded this species as well (see Introduction). The locality of the original is unknown, but Kellogg (1936A) suspects it came from the vicinity of Washington Old Court House, now almost on the line between Washington and Choctaw counties, Alabama. Leidy's (1852) material is from Louisiana.

The vertebrae are intermediate between those of *Basilosaurus* and *Zygorhiza,* not tremendously elongated, but obviously with the centra longer than high. It is a very poorly known species, and its position is uncertain. It may even represent a third family of Alabama archaeocetes.

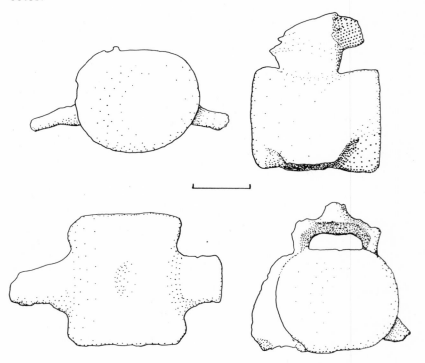

FIG. 87.   *Pontogeneus brachyspondylus* (Müller). Two vertebrae, Choctaw County, Alabama. Left: twelfth lumbar, USNM 2211, anterior and ventral views. Right: third lumbar, USNM 776, lateral and anterior views. After Kellogg, 1936. Scale bar 10 cm.

Suborder Odontoceti
toothed whales

Very little is known of toothed whales in Alabama. The single Eocene specimen, a fragmentary tooth, on which Leidy (1873B, p. 337, pl. 37, fig. 15) based his *Pontobasileus tuberculatus,* only *may* have been from Alabama. This tooth, the only specimen known of the species, has an extremely ornate surface. Kellogg (1936A, p. 264) suggested that this tooth perhaps represented a squalodontid porpoise (a very primitive family of toothed whales).

Whitmore (*in* Isphording and Lamb, 1971) reports *Potamodelphis inaequalis* Allen (1921A, p. 148, pls. 10, 11; see also Kellogg, 1924C) from the Pliocene Citronelle Formation of northern Mobile County, Alabama. This is a member of the Family Iniidae and is closely related to the living Amazon River dolphin *(Inia).* No description of the Alabama material is available.

The odontocetes differ from the archaeocetes in the structure of the skull roof, which is extremely peculiar. The anterior bones of the skull (premaxillae, nasals, maxillae) make up the bulk of the roof, pushing far back and carrying the nostrils with them. They become twisted in the process, so that everything seems out of balance. The bones of the extreme back of the skull also push forward, even to the point of meeting the anterior bones. The traditional bones of the skull roof, the frontals and parietals, are pushed to one side and almost forgotten.

Suborder Mysticeti
whalebone whales

Most of the largest whales living today are in this group; only the sperm whale among the toothed whales attains a comparable size. There are never teeth in the upper jaw, and the lower teeth are retained in the adult only in very primitive forms. They show the same "telescoping" of the skull roof as the odontocetes but in different fashion. There is never any significant "twisting"; everything stays nicely symmetrical. Two blowholes (nostrils) are retained, rather than one as in the toothed whales. As in the odontocetes, the nostrils are far back, on top of the skull. Feeding is accomplished by the use of baleen ("whalebone"), great frilly outgrowths of the ridges on the roof of the mouth. The baleen strains out small marine animals, which are licked off and swallowed.

FAMILY PATRIOCETIDAE

This poorly known family of primitive whales is from the late Eocene and Oligocene. Romer (1966, p. 300) places the family in the Mysticeti, as the pattern of "telescoping" in the roof of the skull resembles the symmetry of the whalebone whales more than the twisted skulls of the toothed whales. He considers the family a possible precursor of the Mysticeti.

?*Archaeodelphis patrius* Allen (fig. 88)

Archaeodelphis patrius Allen, 1921, pp. 2–14, pl. 1, figs. 1–2.

Our question mark on this species does not indicate doubt as to its relationships, nor to the specimen itself (only the type is known!). Our doubts are as to the source of the specimen. It is only speculatively from Alabama, as there was no label. The matrix was examined for microfossils; these indicated a late Eocene age and a locality in the Gulf Coastal Plain, perhaps in Alabama. At least, no other speculations have been presented.

The specimen itself would be all but incomprehensible to anyone but an expert on fossil whales. Nothing is preserved but most of the braincase with small parts of the front of the skull. It represented a small whale, only 5 to 6 feet long, and is roughly fist-sized.

The specimen is in the Museum of Comparative Zoology at Harvard (#15749).

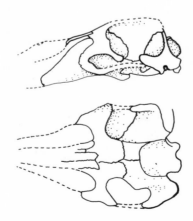

FIG. 88.   ?*Archaeodelphis patrius* Allen. Posterior part of skull, lateral and dorsal views. Approximately x¼. The pattern of the skull roof suggests relations to the ancestry of the baleen whales. After Gregory, from Allen.

Epilogue

We began this discussion by claiming that the fossil vertebrates of Alabama were very poorly known. Yet we have since presented hosts of pages and figures, all to describe what must seem by now a veritable horde of species. Our figures showed many very neat restorations that should have indicated very detailed knowledge of these forms. Are the fossil vertebrates of Alabama really that poorly known?

They are. Many of the species described herein are known from but a single specimen, usually fragmentary. Only a few, some of the sharks, a few bony fishes, several of the turtles, and the archaeocetes, are really well known. Almost all of those neat restorations really should have numerous holes in them representing parts of the skeleton that are totally unknown. Many are based on material from outside Alabama, "imported" to fill in our meager knowledge and make an incomprehensible bit of bone into an animal that you could believe once lived. There is not a single fossil species in Alabama that is so well known that further collecting could not yield valuable new information.

Much information is needed beyond this. There are vast holes in our knowledge that are unfilled by a single specimen. These voids represent animals that we feel certain are around somewhere. But no one has uncovered their bones or teeth and realized what they were. One of the most glaring examples is found in rocks of Pennsylvanian age. *Something* made those tracks that we discussed earlier—something that lived, breathed, ate, and bred—something that died and left bones somewhere. We want to know where!

Out there, in the fields, forests, and streams of Alabama, are the answers to most or all of the questions we have asked. Even more numerous are the answers to questions we haven't even dreamed of asking yet. You won't really need to dig; much of it is there waiting to be picked up.

Please go out and find a new answer. Then join us in working out exactly what that answer is.

Appendix I

Update on Fossil Reptiles of Alabama
Samuel W. Shannon
John T. Thurmond

In the time required to compile this volume, work on Alabama fossils has not been idle. While many finds have been made, and much significant information gained, the largest gains have been in our knowledge of Cretaceous reptiles. Incorporation in the text of this new data would have resulted in even more delays, so we take this opportunity to add an appendix.

Subclass Synaptosauria

Order Sauropterygia

Suborder Plesiosauria

Until 1974, Alabama plesiosaurs were known only from the Mooreville Chalk. Shannon (1974) reports two vertebrae from the Tombigbee Sand Member of the Eutaw Formation, immediately underlying the Mooreville Chalk. These are referable to the Elasmosauridae, but no closer referral can be attempted.

A limb bone fragment from the Bluffport Marl Member of the Demopolis Chalk (younger than the Mooreville) appeared to Welles (letter to Shannon, 1977) to be close to the femur of *Alzadasaurus pembertoni* from the Campanian of South Dakota. This merely means that it is similar to other late plesiosaur material from North America, and represents a time when this once-mighty group was declining to its extinction. The specimen from the Bluffport is one of the two youngest specimens of plesiosaurs from North America.

Subclass Lepidosauria

Order Squamata

Suborder Lacertilia

FAMILY MOSASAURIDAE

Again, recent work has expanded our knowledge of Alabama mosasaurs from the fauna of the Mooreville Chalk only to other units. Shannon (1975) noted specimens from the Tombigbee Sand Member of the Eutaw Formation, the Arcola Limestone Member at the top of the Mooreville Chalk, two members of the Demopolis Chalk, and the Prairie Bluff Chalk. Shannon (1977) has also noted a specimen from the Ripley Formation. Thus, all outcropping Cretaceous formations in Alabama now have produced mosasaur remains except for the Tuscaloosa Formation. This unit, slightly older than those known to contain North American mosasaur remains, is largely continental in character and is not ecologically likely to contain mosasaurs. Several forms have been reported that had not previously been described from Alabama.

Subfamily Mosasaurinae

Mosasaurus sp. cf. *M. missouriensis* (Harlan)

Ichthyosaurus missouriensis Harlan, 1834, p. 440.
Mosasaurus missouriensis, Leidy, 1858, p. 90; Russell, 1967, pp. 136–138.
Mosasaurus sp. cf. *M. missouriensis,* Shannon, 1975, pp. 48–51.

The specimen (GSA-V-1081) was collected by Charles Copeland from the lower Demopolis Chalk in Greene County, Alabama. Fourteen vertebrae and several skull elements are preserved. The articular faces of the vertebrae are circular, or nearly so, and have weakly developed zygosphene-zygantrum articulations above the neural canal.

The tooth crowns are sharply recurved near the tip and bear a pronounced carina. The inner and outer faces of the teeth are somewhat faceted. For descriptions of the skull fragments, see Shannon, 1975.

The genus had not been reported previously from Alabama, though its wide distribution in both time and space suggested that finding it was simply a matter of time. Besides the vertebral characters,

Mosasaurus can easily be distinguished by the combination of the following three skull characters:
1. Sharply pointed, rather slender teeth.
2. Lack of a broad table on the dorsal surface of the parietals between the temporal openings.
3. A large coronoid bone, deeply curved in lateral view.

Only one other specimen, to our knowledge, has been found in Alabama. (See mosasaur discussion in text.)

Subfamily Plioplatecarpinae

Platecarpus sp. cf. *P. somenensis* Thevenin

Platecarpus somenensis Thevenin, 1896, pp. 907–908, pl. 30, fig. 6.
Platecarpus sp. cf. *P. somenensis*, Russell, 1967, pp. 155–156; Thurmond, 1969, pp. 75–76; Shannon, 1975, pp. 54–59, pls. 3–4.

A specimen consisting of a disarticulated skull, a scapula, and fifteen vertebrae was collected by Shannon in the basal part of the Demopolis Chalk in Greene County, Alabama (GSATC-220).

The large size of the dentary (394 millimeters long) characterizes this species, which is also known from France, South Dakota, and Texas. Its trans-Atlantic distribution probably reflects the much narrower Atlantic during Late Cretaceous time than at present.

Subfamily Tylosaurinae

Tylosaurus proriger (Cope)

Macrosaurus proriger Cope, 1869, p. 123.
Liodon proriger, Cope, 1869–70, p. 201.
Tylosaurus proriger, Marsh, 1872, p. 147; Russell, 1967, pp. 173–175, figs. 2c, 5a, 21, 24c, 27, 48a, 55, 63, 92, 93b, 94a, and pl. 1, fig. 3, pl. 2, fig. 3; Thurmond, 1969, pp. 73–74; Shannon, 1975, pp. 69–72, fig. 16b.

In the body of this work, we stated a supposition that this species existed in Alabama, based on specimens in the collections of Birmingham-Southern College. Further work by Shannon (1975) on these specimens indicates that this suspicion is correct. Two individuals are represented, one from near Greensboro, Hale County, and the other from near Fort Deposit, Lowndes County, both in Alabama.

It is impossible to separate these specimens now because someone years ago tossed both into the same box. The Hale County specimen is from the lower part of the Mooreville Chalk, while the Lowndes County specimen represents a later, but indeterminable, Cretaceous unit. These specimens are now in the Red Mountain Museum, Birmingham.

Teeth in place in the maxilla fragments show definite serrations on the anterior carina, a character Thurmond (1969) reported for late specimens of *Tylosaurus* from Texas. The one preserved premaxilla is a massive structure bearing the characteristic "ram" of the genus. The splenial and angular (three of each are preserved, demonstrating the existence of two individuals) definitely lack the distinctive "peg-and-socket" joint that probably belongs to *Tylosaurus zangerli* (Russell, 1970) and are thus referred to the more "normal" *T. proriger*.

From the size, age, and presence of serrations on the teeth, these specimens fit well with other Campanian tylosaurs from North America, some of which attained skull lengths of more than 2 meters (Thurmond, 1969). Further study may demonstrate that these characters are consistent enough to warrant the establishment of a new species.

The vertebrae of *Tylosaurus* have a pear-shaped articular surface that is also characteristic of *Platecarpus*. Weathered centra less than 50 millimeters across probably represent *Platecarpus*, while any more than 100 millimeters across definitely belong to *Tylosaurus*. Those intermediate in size may belong to large individuals of *Platecarpus*, especially to the largest species, *P. somenensis*, or to small, perhaps juvenile, *Tylosaurus*. It should also be remembered that tail vertebrae of either genus can be quite small and that anterior dorsal vertebrae are likely to be large.

Of North American tylosaurs, only *T. nepaeolicus* remains unreported from Alabama. This species seems confined to the earlier parts of the Niobrara Chalk in Kansas and adjacent areas, a time span poorly represented by fossiliferous rocks in Alabama.

other mosasaurs

New material of *Globidens* is now appearing with some frequency in Mississippi, and Alabama fossil hunters are urged to look with special care for this strange mosasaur. Such specimens would be of considerable use in determining the relationships between *G. alabamaensis* and a new species being described by Russell (in press) from South Dakota.

Subclass Archosauria

Order Crocodilia

Three new occurrences of fossil crocodiles are known from Alabama since the main body of the text was written.

Price and Adams (1976) reported a single vertebra from the Nanafalia Formation (early Eocene) of Coffee County, Alabama. No attempt seems to have been made to identify this specimen precisely.

A single caudal vertebra (GSA-V-1073) has been found in the Mooreville Chalk of Dallas County, Alabama. While rather long and narrow, with a spinelike neural process, it has not been identified further, and may not be sufficient to do so.

An isolated tooth (Texas Memorial Museum 40096-1) comes from the Cusseta Sand Member of the Demopolis Formation. It is round in section, striated, and quite sturdy, a combination of features which assure that it is not some other reptile or a fish, and is about 35 millimeters long. Both Langston and Parris (letters to Shannon, 1976) find it to resemble known teeth of the widespread Cretaceous crocodile *Goniopholis*.

Order Pterosauria

These are the flying reptiles so well known to anyone who has examined the popular literature on "dinosaurs." The structure of the wings is batlike in that a membrane is stretched between the hand bones. However, a single finger forms the main support strut at the leading edge of the wing rather than having several fingers inserted in the membrane as in the bats. At least some pterosaurs had a hairlike covering on the skin, though there is some evidence that this was not general. Much effort has been expended in trying to determine whether these creatures could truly fly, rather than simply glide and soar, but the question remains moot.

Pterosaurs range in size from the tiny *Rhamphorhynchus* with the body size of a pigeon to a titanic form recently found in the Late Cretaceous of Big Bend, Texas (where they naturally grew bigger) with a wingspan of at least 15 meters (50 feet).

Lawson (1975) in the paper first describing the giant Texas pterosaur, mentions material at the Field Museum in Chicago that comes from the Mooreville Chalk in Dallas County, Alabama. This material is as yet undescribed and only serves to indicate the presence of this fascinating group in Alabama.

REFERENCES

Cope, E. D. 1869. (Remarks on *Holops brevispinus, Ornithotarsus immanis* and *Macrosaurus proriger*). Proc. Acad. Nat. Sci. Philadelphia, vol. 21, p. 123.

————. 1869–1870. Synopsis of the extinct Batrachia, Reptilia, and Aves of North America. Amer. Phil. Soc. Trans., n.s., issued in parts: 1 (1869), pp. 1–105; 2 (1870), pp. 106–235; 3 (1870), pp. i–vii, 236–252.

Harlan, Richard. 1834. On some new species of fossil saurians found in America. Brit. Assoc. Adv. Sci. Rep. 3rd meeting, Cambridge, 1833, 440 pp.

Lawson, D.A. 1975. Pterosaur from the latest Cretaceous of west Texas: Discovery of the largest flying creature. Science, vol. 187, pp. 947–948.

Leidy, Joseph. 1858. List of extinct Vertebrata, the remains of which have been discovered in the region of the Missouri River: With remarks on their geologic age. Proc. Acad. Nat. Sci. Philadelphia, vol. 9, pp. 89–91.

Marsh, O. C. 1872. Note on *Rhinosaurus*. Am. Jour. Sci., 3rd ser., vol. 4, no. 20, p. 147.

Price, R. C., and Adams, F. W. 1976. Reptilian vertebra from the Nanafalia Formation, Coffee County, Alabama (Abs.). Jour. Ala. Acad. Sci., April 1976 meeting, in press.

Russell, D. A. 1967. Systematics and morphology of American mosasaurs. Peabody Mus. Nat. Hist. Bull. 23, pp. vii, 240.

————. 1970. The Vertebrate fauna of the Selma Formation of Alabama. Pt. 7, The Mosasaurs. Fieldiana: Geology Memoirs, vol. 3, no. 7, pp. 367–380.

Shannon, S. W. 1974. Extension of the known range of the Plesiosauria in the Alabama Cretaceous. Southeastern Geology, vol. 15, no. 4, pp. 193–199.

————. 1975. Selected Alabama mosasaurs. Unpublished Master's Thesis, University of Alabama, 89 pp.

————. 1977. The occurrence and stratigraphic distribution of mosasaurs in the Upper Cretaceous of west Alabama (Abs.). Abstracts with Program, Southeastern Section, Geol. Soc. Am., March 1977 meeting, p. 184.

Thevenin, Armand. 1896. Mosasauriens de la Craie Grise de Vaux-Eclusier près de Péronne (Somme). Soc. Geol. France Bull., 3rd ser., vol. 24, no. 2, pp. 900–916.

Thurmond, J. T. 1969. Notes on mosasaurs from Texas. Texas Jour. Sci., vol. 21, no. 1, pp. 69–80.

Appendix II

Summary List of Fossil Vertebrates of Alabama

Pseudocorax affinis (Agassiz)
Family Alopiidae (thresher sharks) 59
  Alopias latidens alabamensis White
Family Scyliorhinidae (cat sharks, some "dogfish") 59
  Scyliorhinus enniskilleni White
Family Carcharhinidae (requiem sharks) 61
  Galeocerdo clarkensis White
  Galeocerdo alabamensis Leriche
  Hemipristis wyattdurhami White
  Physodon secundus (Winkler)
  Aprionodon greyegertoni (White)
  Negaprion gibbesi gilmorei (Leriche)
  Galeorhinus recticonus claibornensis White
  Galeorhinus sp. cf. *G. falconeri* (White)
  Rhizoprionodon sp.
Order Batoidea (sawfishes, skates, rays) 69
  Suborder Pristoidea (sawfishes)
    Family Pristidae 70
      Ischyrhiza mira Leidy
      Pristis sp.
  Suborder Rajoidea (skates)
    Family Rajidae 72
      Raja sp.
  Suborder Myliobatoidea (rays)
    Family Dasyatidae (stingrays, "stingarees") 73
      Hypolophus sp.
      Dasyatis sp.
    Family Myliobatidae (eagle rays) 75
      Myliobatis sp.
      Aetiobatis sp. (spotted eagle rays)
    Family Rhinopteridae (cow-nosed rays) 76
      Rhinoptera sp.
Subclass Holocephali
  Order Chimaeriformes 77
    Suborder Chimaeroidei (chimaeras)
      Family Edaphodontidae 77
        Edaphodon mirificus Leidy
        Edaphodon barberi Applegate
        Edaphodon sp.
  Chondrichthyes *incertae sedis*
    Order Bradyodonti 78
      Family Petalodontidae 78
        ?Petalodus sp.
      Family Psammodontidae 79
        Psammodus sp.

Class Osteichthyes (bony fishes) 79
  Subclass Sarcopterygii (lobe-finned fishes)
  Subclass Actinopterygii (ray-finned fishes)
    Infraclass Chondrostei
      Order Acipenseriformes (sturgeons) 81
        Family Acipenseridae 81
          Propenser hewletti Applegate

Glossary

accessory cusp One of the cusps on a tooth, especially of a shark, smaller than the main cusp. See also **lateral cusp.**

anterior Pertaining to the front parts of an animal. In human anatomy, pertaining to the abdominal region or ventral side.—**anteriorily** adv.

arch Any V- or Y-shaped process that covers a canal. Usually used with vertebrae. Compare **haemal arch, neural arch.**

articulated skeleton A skeleton preserved with the bones largely still in correct relationship to each other.

atlas The first cervical (neck) vertebra. So called because in human beings it bears the "world" of the skull, as did the mythical Titan of the same name.

axis The second cervical (neck) vertebra.

basal Pertaining to the "lower" surface of a tooth. Compare **occlusal.**

branchial Pertaining to the gills.

buccal Of or close to the lip; "outer": *buccal face of a tooth.* Compare **lingual.**

calcified Reinforced with calcium carbonate or phosphate, but not true bone. Usually used of cartilage.

cartilage Tough, whitish connective tissue.

caudal Of or toward the tail; posterior.

centrum (pl. centra) The body of a vertebra, as opposed to its arches (haemal or neural) and processes.

cephalad Toward the head; anterior.

cervical Pertaining to the neck.

chevron bone The Y-shaped haemal arch, especially one in the tail of a reptile.

circulus (pl. circuli) A growth line on a scale.

crown The exposed portion of a tooth, usually covered by enamel.

ctenoid scale A simple, thin, bony scale with radii in addition to circuli.

cusp Any point or prominence on a tooth crown, especially on a surface with more than one point.

denticle, denticulation Any small toothlike structure, whether an individual bone or a point on a larger tooth.

dentine The dense, hard material of a tooth, not so shiny as the enamel. Dentine makes up all of the root and the bulk of the crown. Histologically it is divided into **orthodentine,** composed of roughly parallel fibers, and **osteodentine,** with irregular, contorted fibers. **Pallial dentine** is a thin layer of orthodentine enclosing a core of osteodentine.

dermal Pertaining to the skin.

dermal bone Any bone formed in the skin rather than internally in the body. Unlike **endochondral bone,** dermal bones ossify directly.

dermal denticle A toothlike structure on or in the skin. These form the scales of sharks and rays, for example. Also called *placoid scales.*

distal   Pertaining to the end of a structure most distant from the attachment or the center of the body. Compare **proximal.**

dorsal   Pertaining to the back of an animal. This is the upper surface in most animals, other than human beings. Compare **ventral.**

enamel   The hard, shiny outer covering on all or part of a tooth crown. Also, any similar covering on a bone or scale. See **ganoine.**

endochondral bone   A bone formed in the deeper layers of the body. It is preformed in cartilage, then ossified. See **dermal bone.**

fenestra (pl. **fenestrae**)   A "window" or large opening in a bone. See **foramen.**

fontanelle   An opening between two bones that normally fuses. The opening may be usual, as in the shells of many turtles, or may indicate youth, as in mammal skulls. The "soft spot" on a human baby's head is a fontanelle.

foramen (pl. **foramina**)   Any opening, usually small, in a bone.

ganoid scale   A scale composed of inner bony and outer ganoine layers, usually rhombic and thick. A fish having such scales is called a *ganoid fish.*

ganoine   A layer of enamel on a bone or scale. The term is not used for teeth. See **enamel.**

gill arches   The bony arches that support the gills of fishes. Seven are present in most embryos, but the first becomes the jaws, and the second is partly involved in jaw support and in hearing.

gill slit   An exterior opening of a gill. Each slit is separate for each gill in most Chondrichthyes but covered by an operculum in Osteichthyes.

haemal   Pertaining to blood.

haemal arch   An arch on the ventral surface of a tail vertebra that covers a main blood vessel.

haemal spine   A spine on the haemal arch.

heterocercal tail   A sharklike tail, in which the vertebral column enters and supports the longer upper (dorsal) part, while the smaller ventral portion is supported by fin rays. A few very early fishes and the ichthyosaurs have *reversed heterocercal* tails, in which the vertebral support is in the ventral lobe.

holotype   See **type.**

homocercal tail   The symmetrical tail of advanced fishes, in which the vertebral column does not enter the tail at all.

hyoid arch   The second of the recognizable vertebrate gill arches, behind the mandibular. Even in mammals, many important structures in the head/neck region derive from it.

hypotype   See **referred (specimen).**

labial   Of or close to the lip; buccal.

lateral   Of or on the side, either of the animal or of the structure.

lateral cusp   An accessory cusp, especially one of a pair of accessory cusps.

lectotype   See **type.**

lingual   Of or close to the tongue; "inner": *lingual surface of a tooth.* Compare **buccal.**

main cusp   The longest point on a tooth crown. Used especially of shark teeth.

mandibular   Pertaining to the jaws, especially the lower jaw. The *mandibular arch* is the first gill arch, except for a doubtful labial arch.

maxilla (pl. **maxillae**)   The main outer tooth-bearing element of the upper jaw in most vertebrates.

medial, median   On the midline, either of the body or of a structure.

neotype   See **type**.

neural arch   The bony arch on the dorsal part of vertebrae through which the spinal cord passes.

occlusal   Pertaining to the surface of a tooth that meets teeth in the other jaw: *occlusal surface, occlusal view.* Preferred in scientific usage to the unclear (and half-true) "upper." Compare **basal**.

operculum   In bony fishes, the plate on the side of the head covering the gills.

orbit   The bony eye socket.

ornament, ornamentation   Any type of "fancy work" on the surface of a bone or tooth. (These are catchall terms for features of unknown adaptive significance.)

orthodentine   See **dentine**.

osseous   Bony.

osteodentine   See **dentine**.

pallial dentine   See **dentine**.

paratype   See **type**.

pharyngeal   Pertaining to the throat region. *Pharyngeal teeth* are on the gill arches.

placoid scale   See **dermal denticle**.

posterior   Pertaining to the hind or caudal parts of an animal. In human anatomy, pertaining to the dorsal side. Compare *anterior.*—**posteriorly** adv.

process   Any protrusion or extension from the main body of a bone.

proximal   Pertaining to the end of a structure (usually a bone) nearest the point of attachment to the body or nearest the center of the body. Compare **distal**.

radius (pl. **radii**)   A radiating ridge extending from nucleus to margin of a ctenoid scale.

referred (specimen)   Any specimen other than the holotype and paratypes (if any) later placed in a species by the same or another worker. Also, a species placed in a genus of which it is not the type species. Sometimes called *hypotype.*

rhombic   Diamond-shaped.

root   The base of a tooth, embedded either in a bony socket or in soft tissues and usually not covered by enamel.

rostral   Pertaining to the rostrum.

rostrum (pl. **rostra** or **rostrums**)   Any nose or beaklike structure at the anterior end of the body.

rugae (sing. **ruga**)   Large ridges or folds, usually parallel.

scute   A bony or horny plate, lying on or in the skin, with others usually forming a partial or complete armor.

serrate(d)   Sawlike.

serration   A series of notches as in the edge of a tooth, giving a sawlike appearance. This type of edge greatly increases the cutting power of the tooth, much like the serrations on certain kinds of knife blades.

sigmoid, sigmoidal   Having an S-shaped curve, especially in side view.

stria (pl. **striae**), **striation**   A small straight groove, especially one of a series of such parallel grooves.

suture   A fairly rigid, interlocking join between two bones, as in a skull.

symphysis (pl. **symphyses**)   A joining of two or more bones, usually along the midline.

syntype   See **type.**

temporal   Pertaining to the temples, the sides of the skull, especially the posterior part.

thorn   A sharp, enlarged dermal denticle of skates.

topotype   See **type.**

tubercule   A small bump, usually rounded and unpointed.

type   In the broadest sense, any specimen on which a new name (usually a species) has been based. Variously modified:

   holotype   The single specimen on which the species was based. Must be designated as such by the original author unless only a single specimen was available at the time.

   hypotype   See **referred (specimen).**

   lectotype   One of a group of syntypes later designated to serve as a holotype.

   neotype   A replacement for a holotype that has been lost or destroyed.

   paratype   Any specimen other than the holotype examined by an author during the description of a new species and included in that species in the original description.

   syntype   One of a group of specimens used in the original description when a holotype was not designated. This is not according to modern usage. Only a syntype may be designated a lectotype, at which point the other syntypes become paratypes.

   topotype   A specimen from the same locality as the holotype or lectotype.

ventral   Pertaining to the lower surface of an animal or the abdominal region of a human being.—**ventrally** adv.

zygantrum   An excavation in the posterior face of the top of a neural arch to receive a zygosphene.

zygapophysis   In land vertebrates, the main articular surface besides the direct contact of centra between vertebrae. They are processes growing laterally from the bases of the neural arch. The anterior zygapophysis on a vertebra *(prezygapophysis)* always has its articular face directed dorsomedially ("up and in"), while that of the *postzygapophysis* is directed ventrolaterally ("down and out").

zygosphene   A process or one of a pair of processes growing from the anterior surface of the top of a neural arch. With the zygantrum it forms an extra articulation between vertebrae.

Bibliography

References are keyed, where possible, to the Hay bibliographies, 1902 and 1929.

Abel, O. 1906. Über den als Beckengürtel von Zeuglodon beschriebenen Schultergürtel eines Vögels aus dem Eocän von Alabama: Centralblatt für Mineral. Geol. und Palaont., Stuttgart, no. 15, pp. 450–458, 4 figs.

Agassiz, L. 1833–1845. Recherches sur les poissons fossiles. Neuchatel, 5 vols., supp., 1420 pp., 396 pls.

————. 1843B. Recherches sur les poissons fossiles. Vol. 1, xxxii + 188 pp., atlas of 11 pls. (A–K). Vol. 2, xii + 310 pp. Vol. 3, viii + 390 + 32 pp., atlas of 83 pls. Vol. 4, xvi + 292 + 22 pp., atlas of 61 pls. Vol. 5, xii + 160 + 144 pp., atlas of 95 pls. Dates from 1833 to 1844.

————. 1845A. Rapport sur les poissons fossiles de l'argile de Londres.

Albritton, C. C., Jr., ed. 1963. The fabric of geology. Reading, Mass.: Addison-Wesley Pub. Co., x + 372 pp., illus.

————. 1967. Uniformity and simplicity: A symposium on the principle of uniformity in nature. Geol. Soc. Am. Spec. Paper 89, [vi] + 99 pp., illus.

Aldrich, T. H., Sr., and Jones, W. B. 1930. Footprints from the Coal Measures of Alabama. Ala. Mus. Nat. Hist., Museum Paper 9, pp. 1–69, pls. 1–17.

Allen, G. M. 1921. A new fossil cetacean. Bull. Mus. Comp. Zool., vol. 65, no. 1, pp. 1–14, 1 pl., 2 figs.

————. 1921A. Fossil cetaceans of the Florida phosphate beds. Jour. Mamm., vol. 2, pp. 104–159, pls. 9–11.

————. 1926. Fossil mammals from South Carolina. Bull. Mus. Comp. Zool., Harvard, vol. 67, no. 14, pp. 447–467, pls. 1–5.

Andrews, C. W. 1910–1913. A descriptive catalogue of the marine reptiles of the Oxford Clay. 2 vols., London, 411 pp., 23 pls.

Anonymous. 1935BO. Giant turtle and mosasaur found in Alabama. Science, n.s., vol. 82, no. 2120, supp., p. 7.

Applegate, S. P. 1970. The vertebrate fauna of the Selma Formation in Alabama. Pt. 8, The fishes. Fieldiana: Geology Memoirs, vol. 3, no. 8, pp. 385–433, figs. 174–204.

Arambourg, C. 1935F. Note preliminaire sur les vertébrés fossiles des phosphates du Maroc. Bull. Soc. Geol. France, ser. 5, vol. 5, pp. 413–439, 2 figs., 2 pls.

————. 1952D. Les vertébrés fossiles des gisements de phosphates (Maroc-Algerie-Tunisie). Notes Mem. Serv. Geol. Maroc. 92, 372 pp., 62 figs., 46 pls., 7 tabs. Appendix by Bergounioux, pp. 375–376.

Bardack, D. 1965. Anatomy and evolution of Chirocentrid fishes. Kans. Univ. Paleo. Cont.: Vertebrata, Art. 10.

Bentley, R. D., Neathery, T. L., and Lines, G. C. 1970. A probable impact-type structure near Wetumpka, Alabama (abs.). Am. Geophys. Union. Trans., vol. 51, no. 4, p. 342.

Berg, Leo Semenovich. 1947. Losses among ichthyologists during the war time. Priroda, vol. 1, p. 95, (in Russian).

———. 1958. System der Rezentens und Fossilen Tischartegen und Fische. Veb. Deutscher der Wissenschaften, Berlin.

Bigelow, B., and Schroeder, W. C. 1948. Fishes of the western North Atlantic. Pt. 1, Mem. Sears Foundation for Marine Research, no. 1, chap. 3, Sharks, pp. 59–576, figs. 6–106.

Boni, Alfredo. 1937B. Il *Lepidopus brevicauda* von Rath del Musco de Pavia. Palaeontogr. Ital., vol. 37, pp. 211–224, 2 pls.

Buckley, S. B. 1843A. Notice of the discovery of a nearly complete skeleton of the *Zeuglodon* of Owen (*Basilosaurus* of Harlan). Am. Jour. Sci., vol. 44, pp. 409–412.

———. 1846A. On the Zeuglodon remains of Alabama. Am. Jour. Sci., ser. 2, vol. 2, pp. 125–129.

Carr, Archie F. 1955. The windward road. Alfred A. Knopf.

———. 1967. So excellent a fishe: A natural history of sea turtles. Doubleday.

Carus, C. G. 1849. Das Kopfskelet des *Zeuglodon hydrarchos*. Nova Acta Acad. Caes. Leop.-Carol. Nat. Cur., Bratislaviae et Bonnae, vol. 22, pp. 2, 371–390, pls. 39a–396b.

———, Geinitz, H. B., Günther, A. F., and Reichenback, H. G. L. 1847. Resultate geologischer, anatomischer, und zoologischer Untersuchungen über das unter dem Namen *Hydrarchos* von Dr. A. C. Koch zuerst nach Europa gebrachte und in Dresden aufgestellte grosse fossile Skelett. Dresden and Leipzig, pp. 1–15, pls. 1–7.

Casier, E. M. 1944A. Contributions à l'étude des poissons fossiles de la Belgique. V: Les genre *Trichiurides* Winkler (s. str.) et *Eutrichiurides* nov. Leurs affinités respectives. VI: Sur le *Sphyraenodus* de l'éocène et sur la presence d'un Sphyraenide dans le Bruxellien (lutetien inférieur). Bull. Mus. Hist. Nat. Belg. 20, no. 11, pp. 1–16, 1 pl.

———. 1946. La faune ichthyologique de l'ypresien de la Belgique. Mem. Mus. Hist. Nat. Belg. 104, pp. 3–267, 19 figs., 6 pls.

———. 1947. Constitution et évolution de la racine dentaire des Euselachii [3 pts.]. Bull. Mus. Roy. Hist. Nat. Belg., vol. 23, nos. 13, 14, 15.

———. 1953. Origine des Ptychodontes. Mem. Inst. Roy. Sci. Nat. Belg., ser. 2, no. 49.

———. 1958. Contribution à l'étude des poissons fossiles des Antilles. Schweiz. Pal. Abl., no. 74, 95 pp., 7 figs., 3 pls., 9 tabs.

———. 1960. Notes sur la collection des poissons Paleocenes et Éocènes de l'Enclave deCabinda (Congo). Ann. Mus. Roy. Cong. Belg., Terouren, ser. 3, vol. 1, fasc. 2, pp. 1–48, 7 figs., 2 pls.

Clark, W. B. 1895A. Contributions to the Eocene fauna of the middle Atlantic slope. Johns Hopkins Univ. Circ. 15, pp. 3–6.

———. 1897A. Eocene deposits of middle Atlantic slope. U.S. Geol. Surv. Bull. 141, pp. i–vii, 1–167, pls. 1–11.

Claypole, E. W. 1894A. *Cladodus? magnificus*, a new Selachian. Am. Geol., vol. 14, pp. 137–140, pl. 5.

Collins, R. E. L. 1954. A new turtle, *Toxochelys weeksi*, from the Upper Cretaceous of west Tennessee. Jour. Tenn. Acad. Sci., vol. 26, pp. 262–269, 2 pls.

218   Bibliography

Cope, E. D. 1869A. On the reptilian orders Pythonomorpha and Streptosauria. Proc. Boston Soc. Nat. Hist., vol. 12, pp. 250–266.

———. 1869B. On some Cretaceous reptilia. Proc. Acad. Nat. Sci. Philadelphia, vol. 20, pp. 233–242.

———. 1869D. Remarks on fossil reptiles, *Clidastes propython, Polycotylus latipinnis, Ornithotarsus immanis.* Proc. Am. Phil. Soc., vol. 11, p. 117.

———. 1869K. The fossil reptiles of New Jersey. Am Nat., pt. 1, vol. 1, pp. 23–30; pt. 2, vol. 3, pp. 84–91, pl. 2.

———. 1869L. Description of some extinct fishes previously unknown. Proc. Boston Soc. Nat. Hist., vol. 12, pp. 310–317.

———. 1869M. Synopsis of the extinct *Batrachia, Reptilia,* and *Aves* of North America. Trans. Am. Phil. Soc., vol. 14, pp. i–viii, 1–252, pls. 1–14, 55 woodcuts.

———. 1870F. Fourth contribution to the history of the fauna of the Miocene and Eocene periods of the United States. Proc. Am. Phil. Soc., vol. 11, pp. 285–294.

———. 1870L. On the Saurodontidae. Proc. Am. Phil. Soc., vol. 11, pp. 529–538.

———. 1871A. [Verbal communication on *Pythonomorpha*]. Proc. Am. Phil. Soc., vol. 11, pp. 571–572.

———. 1871N. Catalogue of the *Pythonomorpha* found in the Cretaceous strata of Kansas. Proc. Am. Phil. Soc., vol. 12, pp. 264–287.

———. 1872EE. Remarks on geology of Wyoming, and on saurodont fishes from Kansas. Proc. Acad. Nat. Sci. Philadelphia, 1872, pp. 279–281.

———. 1872I. On the families of fishes of the Cretaceous formation of Kansas. Proc. Am. Phil. Soc., vol. 12, pp. 327–357.

———. 1872L. A description of the genus *Protostega,* a form of extinct *Testudinata.* Proc. Am. Phil. Soc., vol. 12, pp. 422–433.

———. 1874C. Review of the *Vertebrata* of the Cretaceous period found west of the Mississippi River. Sec. I, On the mutual relations of the Cretaceous and Tertiary formations of the West. Sec. II, List of species of *Vertebrata* from the Cretaceous formations of the West. Bull. U.S. Geol. & Geog. Surv. Terr., vol. 1, no. 2, 1874, pp. 3–48.

———. 1875E. The *Vertebrata* of the Cretaceous formations of the West. Rept. U.S. Geol. Surv. Terr., vol. 2, pp. 1–303, pls. 1–55, 10 woodcuts. Washington, D.C., 1875.

———. 1875V. Synopsis of the *Vertebrata* whose remains have been preserved in the formations of North Carolina. Rept. Geol. Surv. of North Carolina 1, by W. C. Kerr, Appendix B, pp. 29–52, pls. 5–8.

———. 1876F. On a new genus of fossil fishes. Proc. Acad. Nat. Sci. Philadelphia, 1876, p. 113.

———. 1877P. On a new species of *Adocidae* from the Tertiary of Georgia. Proc. Am. Phil. Soc., vol. 17, pp. 82–84.

———. 1878EE. Paleontology of Georgia. Am. Nat., vol. 12, p. 128.

———. 1882J. The reptiles of the American Eocene. Am. Nat., vol. 16, pp. 979–993.

———. 1890A. The Cetacea. Am. Nat., vol. 24, no. 283, pp. 599–616, pls. 20–23, 8 text-figs.

―――. 1890F. The extinct sirenia. Am. Nat., vol. 24, pp. 697–702, pl. 26, 3 figs.

―――. 1893A. A preliminary report on the vertebrate paleontology of the Llano Estacado. Geol. Surv. Texas, 4th Ann. Rept., pp. 1–136, pls. 1–23.

Copeland, C. W., Jr. 1963. Curious creatures in Alabama rocks: A guide book for amateur fossil collectors. Geol. Surv. Ala. Circ. 19, pp. 1–45, 28 figs., 1 pl.

Dames, W. 1883A. Über *Ancistrodon* Debey. Zeitschr. deutsch. geol. Gesellsch., vol. 35, pp. 655–670, pl. 19.

―――. 1883B. Über eine tertiäre Wirbelthierfauna von der westlichen Insel der Birket-el-Qurun im Fajcim (Aegypten). Sitzungsber. k. Preuss. Akad. Wiss., Berlin, 1883, pp. 6, 129–133.

―――. 1894A. Die Chelonier der norddeutschen Tertiärformation. Palaeontolog. Abhandl., vi. (V in Hay, 1908), p. 220.

Darteville, E., and Casier, E. 1959. Les poissons fossiles du Bas-Congo et des regions voisins. Pt. 3, Ann. Mus. Congo Belge, A, ser. 3, vol. 2, pp. 257–268, figs. 77–98, pls. 23–39, tabs. 1–18.

Dean, Bashford. 1909C. Studies on fossil fishes (sharks, chimaeroids, and arthrodires). Am. Mus. Nat. Hist. Mem. 9, pp. 211–287, pls. 26–41, 65 text-figs.

DeKay, J. G. 1842. Zoology of New York. Pt. 1, Mammalia. Albany, xiii + 146 pp., 33 pls.

Desmarest, A. G. 1822A. Mammalogie ou description des espèces de mammifères. 4 vols., vii + 556 pp., 126 pls.

Dollo, L. 1924. *Globidens alabamaensis*, mosasaurien americain retrouve dans le Craie d'Obourg (Senonien superieur) du Hainaut, et les mosasauriens de la Belgique en general. Arch. Biol., vol. 34, pp. 167–213.

Dowling, H. G., Jr. 1941. A new mosasaur skeleton from the Cretaceous in Alabama. Ala. Acad. Sci. Jour., vol. 13, pp. 46–48, 1 fig.

Drevermann, F. 1933. Das Skelett von *Placodus gigas* im Senckenberg-Museum. Abh. Senck. nat. Gesellsch., vol. 38, pp. 319–364.

Dunkle, D. H. 1958A. Three North American Cretaceous fishes. Proc. U.S. Nat. Mus. 108, pp. 269–277, 3 pls.

Eastman, C. R. 1895A. Beiträge zur Kenntniss der Gattung *Oxyrhina*, mit besonderer Berücksichtigung von *Oxyrhina mantelli* Ag. Palaeontographica, vol. 6, pp. 149–191, pls. 16–18.

―――. 1901. *Pisces*, pp. 95–115, and pls. 12–13 in Maryland Geol. Surv., Eocene, pp. 1–332, pls. 1–64.

―――. 1906E. Sharks' teeth and cetacean bones. Bull. Mus. Comp. Zool., vol. 50, pp. 74–98, pls. 1–3, 7 figs.

―――. 1907. Types of fossil cetaceans in the Museum of Comparative Zoology, Harvard College. Bull. Mus. Comp. Zool., vol. 51, no. 3, pp. 79–94.

―――. 1917A. Fossil fishes in the collection of the United States National Museum. U.S. Nat. Mus. Proc., vol. 52, pp. 235–304, pls. 1–23, 9 tabs.

Eckert, A. W. 1963. Coon Creek's fabulous fossils. Sci. Digest, vol. 53, pp. 47–53, 7 figs.

Emmons, E. 1845A. On the supposed *Zeuglodon cetoides* of Professor Owen. Am. Quart. Jour. Agri. & Sci., vol. 2, pp. 59–63, 366.

——. 1846A. Description of some of the bones of the *Zeuglodon cetoides* of Professor Owen. Am. Quart. Jour. Agri. & Sci., vol. 3, pp. 223–231, pls. 1, 2.

——. 1858B. Report of the North Carolina Geological Survey: Agriculture of the eastern counties, together with descriptions of the fossils of the marl beds. Raleigh, xv + 314 pp.

Estes, R. 1964. Fossil vertebrates from the Late Cretaceous Lance Formation, eastern Wyoming. Univ. Calif. Publ. Geol. Sci., vol. 49, pp. 1–187, 5 pls., 73 figs., 7 tabs.

Falconer, H. 1857B. On the species of mastodon and elephant occurring in the fossil state in Great Britain. Pt. 1, Mastodon. Quart. Jour. Geol. Soc. London, vol. 13, pp. 307–360, pls. 11–12.

——. 1868A. Palaeontological memoirs and notes. Vol. 1, Fauna antiqua sevalensis, pp. 1–590, pls. 1–34, 16 figs. Vol. 2, Mastodon, Elephant, Rhinoceros, etc., pp. 1–675, pls. 1–38, figs. 1–9. London.

Fowler, H. W. 1911A. A description of the fossil fish remains of the Cretaceous, Eocene, and Miocene formations of New Jersey. Bull. Geol. Surv. N.J., vol. 4, pp. i–vi, 22–182, 108 figs.

Fraas, E. 1910. Plesiosaurier aus dem oberen Lias von Holzmaden. Palaeontogr., vol. 57, pp. 1–40.

Frick, Childs. 1937. Horned ruminants of North America. Bull. Am. Mus. Nat. Hist., vol. 69.

Frizzell, D. L., and Lamber, C. K. 1961. New genera and species of myripristid fishes, in the Gulf Coast Cenozoic, known from otoliths (Pisces, Beryciformes). Bull. Univ. Mo. School Mines, Met., Tech. Ser. no. 100, pp. 1–25, 22 figs.

——. 1962. Distinctive "congrid type" fish otoliths from the Lower Tertiary of the Gulf Coast (Pisces Anguilliformes). Proc. Calif. Acad. Sci., vol. 32, pp. 87–101, 12 figs.

Gaffney, E. S., and Zangerl, R. 1968. A revision of the Chelonian genus *Bothremys* (Pleurodira: Pelomedusidae). Fieldiana: Geology, vol. 16, no. 7, pp. 193–239, 22 figs.

Gibbes, R. W. 1845A. Description of the teeth of a new fossil animal found in the Green-sand of South Carolina. Proc. Acad. Nat. Sci. Philadelphia, 1845, pp. 254–256, pl. 1.

——. 1846A. On the fossil *Squalidae* of the United States. Proc. Acad. Nat. Sci. Philadelphia, vol. 3, pp. 41–43.

——. 1847. On the fossil genus *Basilosaurus*, Harlan (*Zeuglodon*, Owen), with a notice of specimens from the Eocene Green-sand of South Carolina. Jour. Acad. Nat. Sci. Philadelphia, ser. 2, vol. 1, pp. 2–15, 5 pls.

——. 1848B. Monograph of the fossil *Squalidae*, of the United States. Jour. Acad. Nat. Sci. Philadelphia, vol. 1, pp. 191–206, pls. 18–21.

——. 1849A. Monograph of the fossil *Squalidae*, of the United States. Jour. Acad. Nat. Sci. Philadelphia, vol. 2, pp. 191–206, pls. 25–27.

——. 1850F. Fossils common to several formations. Proc. Am. Assoc. Adv. Sci., 3rd meeting, Charleston, S.C., 1850, pp. 70–71.

——. 1850G. New species of *Myliobates*, from the Eocene of South Carolina, with other genera not heretofore observed in the United States. Jour. Acad. Nat. Sci. Philadelphia, ser. 2, vol. 1, pp. 299–300.

————. 1851. A memoir on *Mosasaurus* and three allied new genera. Smiths. Contr. Knowl., vol. 2, no. 5, pp. 1–13.

Gidley, J. W. 1901A. Tooth characters and revision of the North American species of the genus *Equus*. Bull. Am. Mus. Nat. Hist., vol. 14, pp. 91–142, pls. 18–21, 27 text-figs.

————. 1926F. Description of fishes from the Ripley Formation, Tennessee. U.S. Geol. Surv. Prof. Paper 137, p. 192, pl. 71.

Gilmore, C. W. 1912. A new mosasauroid reptile from the Cretaceous of Alabama. Proc. U.S. Nat. Mus., vol. 41, pp. 470–484, pls. 39, 40, 3 text-figs.

————. 1919C. New fossil turtle, with notes on two described species. Proc. U.S. Nat. Mus., vol. 56, pp. 113–132, pls. 29–37.

————. 1927. Note on a second occurrence of the mosasaurian reptile *Globidens*. Science, n.s., vol. 66, p. 452.

————. 1928. Fossil lizards of North America. Mem. Nat. Acad. Sci., vol. 22, pp. 1–201.

Glykmon, L. S. 1964. Akuly paleogena i ikh stratigraficheskoe znachenie. Acad. Sci., USSR, Moscow.

Goin, C. J., and Auffenberg, W. 1958. New salamanders of the family Sirenidae from the Cretaceous of North America. Fieldiana: Geology, vol. 10, pp. 449–459, figs. 187–189.

Gregory, J. T. 1950. A large pycnodont from the Niobrara Chalk. Postilla, no. 5, pp. 1–10, 2 figs.

Harlan, R. 1824A. On a new fossil genus of the order *Enalio Sauri* (of Conybeare). Jour. Acad. Nat. Sci. Philadelphia, vol. 3, pp. 331–337, pl. 12, figs. 1–15.

————. 1834C. Notice of fossil bones found in the Tertiary formation of the State of Louisiana. Trans. Am. Phil. Soc., n.s., vol. 4, pp. 397–403, pl. 20, figs. 1–3.

————. 1835. Medical and physical researches: Or original memoirs in medicine, surgery, physiology, geology, zoology and comparative anatomy. Philadelphia, Lydia R. Bailey, *c.* 56 pp.

Hatcher, J. B. 1901A. Some new and little known fossil vertebrates. Annals Carnegie Mus., p. 128, pl. 1, figs. 5–6.

Hay, O. P. 1899D. Descriptions of two new species of tortoises from the Tertiary of North America. Proc. U.S. Nat. Mus., vol. 22, pp. 22–24, pls. 4–6.

————. 1899E. On some changes in the names, generic and specific, of certain fossil fishes. Am. Nat., vol. 33, pp. 783–792.

————. 1902. Bibliography and catalogue of the fossil vertebra of North America. U.S. Geol. Sur. Bull. 179, 868 pp.

————. 1903A. On certain genera and species of North American Cretaceous actinopterous fishes. Bull. Am. Mus. Nat. Hist., vol. 19, pp. 1–95, pls. 1–5, 72 text-figs.

————. 1908A. The fossil turtles of North America. Carnegie Inst. Wash. Pub. N. 75, iv + 568 pp., 113 pls., 704 text-figs.

————. 1913B. Notes on some fossil horses, with descriptions of four new species. Proc. U.S. Nat. Mus., vol. 44, pp. 569–594, pls. 69–73, 28 text-figs.

————. 1923. The Pleistocene of North America and its vertebrated animals from the states east of the Mississippi River and from the Canadian prov-

inces east of longitude 95°. Carnegie Inst. Wash. Pub. No. 322, viii + 499
pp., 41 maps, 25 figs.
―――. 1929. Second bibliography and catalogue of the fossil vertebrates of
North America. Carnegie Inst. Wash. Pub. No. 39, 1, vii + 916 pp.
Hays, I. 1830A. Description of a fragment of the head of a new fossil animal
discovered in a marl pit near Moorestown, N.J. Trans. Am. Phil. Soc., ser. 2,
vol. 3, pp. 471–477, pl. 16.
Hibbard, C. W. 1940. A new Pleistocene fauna from Meade County, Kansas.
Trans. Kans. Acad. Sci., vol. 43, pp. 417–425.
―――. 1944B. A new land tortoise, *Testudo riggsi*, from the Middle Pliocene
of Seward County, Kansas. Bull. Univ. Kans. 30, pp. 71–76, 2 figs.
―――. 1955. The Jinglebob interglacial (Sangamon?) fauna from Kansas and
its climatic significance. Univ. Mich. Mus. Paleont. Contr., vol. 12, no. 10,
pp. 179–228, 2 pls., 8 figs., 1 chart.
―――, and Taylor, D. W. 1960. Two late Pleistocene faunas from southwestern Kansas. Univ. Mich. Mus. Paleont. Contr., vol. 16, no. 1, pp. 1–223, 16
pls., 18 figs.
Hopkins, F. V. 1871A. Second Annual Report of the Geological Survey of
Louisiana to the General Assembly. Rept. Louisiana Univ. 2, 1870, pp. 1–35.
Hussakof, L. 1908A. Catalogue of types and figured specimens of fossil ver-
tebrates in the American Museum of Natural History. Pt. 1, Fishes. Bull.
Am. Mus. Nat. Hist., vol. 25, pp. 1–103, pls. 1–6, 49 text-figs.
Isphording, W. C., and Lamb, G. M. 1971. Age and origin of the Citronelle
Formation in Alabama. Geol. Soc. Am. Bull., vol. 82, pp. 775–780, 1 fig.
Jaekel, O. 1894A. Die eocänen Selachier vom Monte Bolca. Ein Beitrag zur
Morphogenie der Wirbelthiere. Berlin, 1894, pp. 1–176, 8 pls., 39 text-figs.
Jones, W. B. 1930. Footprints found in Alabama mine. Coal Age, vol. 35, no. 2,
p. 92, 3 figs.
Kauffman, E. G., and Kesling, R. V. 1960. An Upper Cretaceous ammonite
bitten by a mosasaur. Univ. Mich. Mus. Paleont. Contr., vol. 15, no. 9, pp.
193–248.
Kellburg, J. W., and Maher, S. W. 1959. Some fossils from the Maury Formation, DeKalb County, Tennessee. Jour. Tenn. Acad. Sci., vol. 34, pp. 136–
138.
Kellogg, R. 1924C. Tertiary pelagic mammals of eastern North America. Geol.
Soc. Am. Bull., vol. 35, pp. 755–766.
―――. 1928. The history of whales—their adaptation to life in the water.
Quart. Rev. Biol., Baltimore, vol. 3, no. 1, pp. 29–76, figs. 1–11; no. 2, pp.
174–298, figs. 12–24.
―――. 1931A. Ancient relatives of living whales. Explor. Field Work Smiths.
Inst., 1930 (1931), pp. 83–90, 5 figs.
―――. 1936A. A review of the Archaeoceti. Pub. Carnegie Inst. Wash. 482, xv
+ 366 pp., 88 figs., 37 pls.
Kerr, Robert, Linnaeus, Carolus, and Gmelin, Johann Fridrich. 1792. The
animal kingdom or zoological system. Class I (Mammalia), 8 + 28 + 644 pp.,
9 pls.
Koch, A. C. 1845A. Hydrargos, or great sea serpent, of Alabama, etc. New
York, pp. 1–16, 1 fig.

———. 1845B. Description of the Hydrarchos harlani (Koch). Td. ed., New York: B. Owen, printer, 29 Ann St., N.Y., pp. 1–24.

———. 1846. Kurze Beschreibung des *Hydrarchos harlani* (Koch), eines riesen mossegen Meerungeheuers und dessen Entdeckung in Alabama in Nordamerika im Frühjahr 1845. Dresden, pp. 1–20, pl. 1.

Koken, E. 1888A. Neue intersuchungen an tertiären Fische—Otolithen. Zeitschr. deutsch geol. Gesellsch., vol. 11, pp. 274–305, pls. 17–19.

Langston, Wann. 1960. The dinosaurs. Pt. 6 of The vertebrate fauna of the Selma Formation of Alabama. Fieldiana: Geology Memoirs, vol. 3, no. 6, pp. 313–361, figs. 146–163, pl. 34.

Leidy, J. 1851G. Descriptions of a number of fossil reptiles and mammals. Proc. Acad. Nat. Sci. Philadelphia, vol. 5, pp. 325–328.

———. 1852. Description of *Pontogeneus priscus*. Proc. Acad. Nat. Sci. Philadelphia, vol. 6, p. 52.

———. 1855B. A memoir on the extinct sloth tribe of North America. Smiths. Contr. Knowledge, vol. 7, art. 5, pp. 1–68, pls. 1–16.

———. 1855C. Indications of twelve species of fossil fishes. Proc. Acad. Nat. Sci. Philadelphia, 1855, pp. 395–397.

———. 1856A. Description of two ichthyodorulites. Am. Jour. Sci., ser. 2, vol. 21, pp. 421–422.

———. 1856E. Description of two ichthyodorulites. Proc. Acad. Nat. Sci. Philadelphia, vol. 8, pp. 11–12.

———. 1856I. Notices of remains of extinct *Mammalia*, discovered by Dr. F. V. Hayden in Nebraska Territory. Proc. Acad. Nat. Sci. Philadelphia, vol. 8, pp. 88–90.

———. 1856K. Notices of remains of extinct vertebrate animals of New Jersey, collected by Professor Cook, of the state geological survey, under the direction of Dr. W. Kitchell. Proc. Acad. Nat. Sci., Philadelphia, vol. 8, pp. 220–221.

———. 1856O. Remarks on certain extinct species of fishes. Proc. Acad. Nat. Sci. Philadelphia, vol. 8, pp. 301–302.

———. 1856P. Notices of remains of extinct turtles of New Jersey, collected by Professor Cook of the state geological survey, under the direction of Dr. W. Kitchell. Proc. Acad. Nat. Sci. Philadelphia, vol. 8, pp. 303–304.

———. 1857. Notices of some remains of extinct fishes. Proc. Acad. Nat. Sci. Philadelphia, 1857, pp. 167–168.

———. 1858B. Description of *Mastodon mirificus* and *Elephas imperator*. Proc. Acad. Nat. Sci. Philadelphia, 1858, p. 10.

———. 1860A. Extinct *Vertebrata* from the Judith River and great Lignite formations of Nebraska. Trans. Am. Phil. Soc., vol. 11, pp. 139–154, pls. 8–11.

———. 1860B. Description of vertebrate fossils: Holmes's Post-Pliocene fossils of South Carolina. Charleston: Holmes's Book House, pp. 99–122, pls. 15–27.

———. 1865A. Memoir on the extinct reptiles of the Cretaceous formations of the United States. Smiths. Contr. Knowledge, vol. 14, art. 6, pp. 1–135, pls. 1–20.

———. 1868. Notice of American species of *Ptychodus*. Proc. Acad. Nat. Sci. Philadelphia, 1868, pp. 205–208.

———. 1869A. The extinct mammalian fauna of Dakota and Nebraska, including an account of some allied forms from other localities, together with a synopsis of the mammalian remains of North America. Jour. Acad. Nat. Sci. Philadelphia, ser. 2, vol. 7, pp. 1–472, 30 pls.

———. 1870. [A fossil vertebrate from the Cretaceous Formation of Pickens Co., Alabama]. Proc. Acad. Nat. Sci. Philadelphia, 1870, 4, Jan. 1870.

———. 1870E. [Remarks on *Poicilopleuron valens, Clidastes intermedius, Leiodon proriger, Baptemys wyomingensis,* and *Emys stevensonianus*]. Proc. Acad. Nat. Sci. Philadelphia, 1870, pp. 3–5.

———. 1873B. Contributions to the extinct vertebrate fauna of the Western Territories. Rept. U.S. Geol. Surv. Terr., vol. 1, pp. 14–358, pls. 1–37.

———. 1877A. Description of vertebrate remains, chiefly from the phosphate beds of South Carolina. Jour. Acad. Nat. Sci. Philadelphia, ser. 2, vol. 8, pp. 209–261, pls. 30–34.

Leriche, M. 1938C. Quelques observations critiques sur une mémoire de Mlle. W.-A.-E. Vande Geyn, intitulé: "Das Tertiar der Niederlande mit besonderer Berucksichtigung der Selachierfauna." Ann. Soc. Geol. Belg. (8°), 62, pp. B131–B141.

———. 1938D. Contribution à l'étude des poissons fossiles des pays riverains de la Méditerranée americaine (Venezuela, Trinite, Antilles, Mexique). Abh. Schweiz Pal. Geol. 61, pp. 1–41, 8 figs., 4 pls.

———. 1940A. Le synchronisme des formations éocènes, marines, des deux côtés de L'Atlantique, d'après leur faune ichthyologique. Comptes. Rendus Acad. Sci., vol. 210, p. 590.

———. 1940B. Les poissons de la base du Neogene de la Campine et du Petit-Brabant. Ann. Soc. Geol. Belg. 63, 205–216.

———. 1942B. Contribution à l'étude des faunes ichthyologique marines des terrains Tertiaires de la plaine côtière Atlantique et du centre des États-Unis. Mem. Soc. Geol. France, n.s., 20, no. 45, pp. 5–110, 8 figs., 8 pls.

Linnaeus, C. 1758. Systema Naturae, 10th ed., 8 vols., vol. 1, pp. 1–824.

Loomis, F. B. 1900A. Die Anatomie und die Verwandtschaft der Ganoid-und Knochens-Fische aus der Kreide-Formation von Kansas, U.S.A. Palaeontographica 46, pp. 213–283, pls. 19–27, 13 text-figs.

Lucas, F. A. 1898B. A new snake. Proc. U.S. Nat. Mus., vol. 21, pp. 637–638.

Lull, R. S., and Wright, N. 1942. Hadrosaurian dinosaurs of North America. Geol. Soc. Am. Spec. Paper 40, pp. 1–242, 90 figs., 31 pls.

Lundelius, E. L., Jr. 1969. The age structure of a *Tanupolama* sample from South Texas and its ecological significance (abs.). Geol. Soc. Am. Spec. Paper 121, pp. 402–403.

McNulty, C. L., Jr. 1963. Teeth of *Petalodus alleghaniensis* Leidy from the Pennsylvanian of north Texas. Texas Jour. Sci., vol. 15, pp. 351–353, 1 fig.

Maher, S. W., and Dunkle, D. H. 1955. An occurrence of a pleuropterygian shark in the Chattanooga Shale of Tennessee. Jour. Tenn. Acad. Sci., vol. 30, pp. 202–203.

Marsh, O. C. 1870E. Notice of some new Tertiary and Cretaceous fishes. Proc. Am. Assoc. Adv. Sci., 18th meeting, Salem, 1869, pp. 227–230.

————. 1889F. Discovery of Cretaceous *Mammalia*, pt. 2. Amer. Jour. Sci., (3) vol. 138, pp. 177–180, pls. 7–8.

Mercer, H. C. 1897. The finding of the remains of the fossil sloth at Big Bone Cave, Tennessee, in 1896. Proc. Am. Phil. Soc., vol. 36, pp. 36–70, 26 figs.

Morton, S. G. 1830A. Synopsis of the organic remains of the ferruginous sand formation; with geological remarks. Am. Jour. Sci., vol. 17, pp. 274–295; vol. 18, pp. 243–250.

————. 1835A. Notice of the fossil teeth of fishes of the United States, the discovery of the Galt in Alabama, and a proposed division of the American Cretaceous group. Am. Jour. Sci., vol. 28, pp. 276–278.

————. 1842A. Description of some new species of organic remains of the Cretaceous group of the United States; with a tabular view of the fossils hitherto discovered in this formation. Jour. Acad. Sci. Proc., vol. 8, pp. 207–227, pls. 10, 11.

Müller, G. 1847A. *Basilosaurus*. Am. Jour. Sci. (2), vol. 4, pp. 421–422.

————. 1847B. Über die von Herrn Koch in Alabama gesammelten fossilien Knochenreste seines Hydrarchus. Archiv. Anat. Physiol., u. Wiss. Med., vol. 14, pp. 363–377, 378–396.

————. 1847C. Untersuchungen über den Hydrarchos. Ber. k. p. Akad. Wiss. Berlin, 1847, pp. 103–114. Reprint in Neues Jarb. Min., 1847, pp. 623–631.

————. 1847D. Über den Bau des Schädels des *Zeuglodon cetoides* Ow. Ber. k. p. Akad. Wiss. Berlin, 1847, p. 160.

————. 1847E. Über die Wirbelsäule des *Zeuglodon cetoides* Ow. Ber. k. p. Akad. Wiss. Berlin, 1847, p. 160.

————. 1849A. Über die fossilen Reste der Zeuglodonten von Nordamerika, mit Rücksicht auf die europäischen Reste aus dieser Familie. Berlin, pp. i–iv, 1–38, fol., 27 pls.

————. 1851. Neue Beiträge zur Kenntniss der Zeuglodonten. Monatsber. Königl. preuss., Akad. Wiss. Berlin, 28 April 1851, pp. 236–246.

Newberry, J. S. 1889A. The Paleozoic fishes of North America. Monogr. U.S. Geol. Surv., vol. 16, pp. 1–340, pls. 1–53.

————, and Worthen, A. H. 1866A. Descriptions of new species of vertebrates, mainly from the Subcarboniferous limestone and Coal Measures of Illinois. Geol. Surv. Ill., vol. 2, pp. 9–134, pls. 1–13.

Osborn, H. F. 1918A. Equidae of the Oligocene, Miocene, and Pliocene of North America; iconographic type revision. Mem. Am. Mus. Nat. Hist., n.s., vol. 2, pp. 1–330, pls. 1–44, 173 text-figs.

————. 1936–1942. Proboscidea: A monograph of the discovery, evolution, migration and extinction of the mastodonts and elephants of the world. New York, Am. Mus. Nat. Hist. Press, 2 vols., 1675 pp., figs.

Owen, Richard. 1839B. Observations on the teeth of the *Zeuglodon*, *Basilosaurus* of Dr. Harlan. Proc. Geol. Soc. London, vol. 3, pp. 24–28.

————. 1841C. Observations on the *Basilosaurus* of Dr. Harlan (*Zeuglodon cetoides* Owen). Trans. Geol. Soc. London, (2), vol. 6, pp. 69–79, pls. 7–9.

Patterson, Colin. 1966. British Wealden sharks. Bull. Brit. Mus. (Nat. Hist.), Geol., ser. 2, vol. 7, pp. 283–350, 31 figs., 5 pls., 2 tabs.

Pauca, Mircea. 1930A. Revision der fossilen *Lepidopus*-und *Capros*-Arten. Acad. Romana, Bull. Sec. Slun., vol. 13, pp. 177–183.

———. 1934C. Die fossile Fauna und Flora aus dem Olegozan von Suslanesti (Muscel) in Rumänien An. Inst. Geol. Romaniei, vol. 16, pp. 575–668, 30 figs., 7 pls.

Piveteau, J., ed. 1966. Traité de Paléontologie, 3 vols., vol. 4, L'origine des Vertébrés: leur expansion dans les eaux douces et le nailieu marin. Pt. 3. Actinopterygiens, Crossopterygiens, Dipneustes. Lehman G.-P. Paris: Masson et Cie, 440 pp., 357 figs., 11 pls.

Poll, Max. 1951. Poissons. I-Generalites. II-Selaciens et Chimeres. Inst. Royal des Sciences Nat. Belg., Expedition Oceanog. Belg. Eaux Côtières Afr. de L'Atlantique. Résultats Scientifiques, vol. 4, fasc. 1, pp. 1–154, 13 pls., 67 figs.

Quaas, A. 1902A. Beitrag zur Kenntnis der Fauna der obersten Kreidebildungen in der libyishen Wüste (Overwegischichten und Blatterthone). Paleontographica, vol. 30, pt. 2, pp. 153–336, pls. 20–32.

Rapp, W. F. 1944. Check list of the fossil reptiles of New Jersey. Jour. Paleont., vol. 18, pp. 285–288.

———. 1946. Check list of the fossil fishes of New Jersey. Jour. Paleont., vol. 20, pp. 510–513.

Renger, J. J. 1934. Turtle and mosasaur fossils found in the Selma Formation of Greene County, Alabama. Manuscript in Ala. Geol. Surv. Libr., 23 pp.

———. 1934. Prehistoric marine reptiles from the Cretaceous of Alabama. Rocks and Minerals (Peekskill, N.Y.), vol. 9, pp. 109–111, August 1934.

———. 1935A. Excavations of Cretaceous reptiles in Alabama. Sci. Monthly, vol. 41, no. 6, pp. 560–565, 5 figs.

———. 1935B. The fossils of giant reptiles from the Cretaceous of Alabama (abs). Ala. Acad. Sci. Jour., vol. 6, p. 24.

Richards, H. G., and Hand, B. M. 1958. Fossil shark teeth from the Coastal Plain of Georgia. Ga. Min. Newsletter, vol. 11, no. 3, pp. 91–95, 2 pls.

Richards, H. G., et al. 1962. The Cretaceous fossils of New Jersey. Pt. 2, Bull. Bur. Geol. Topog. N.J. Paleont. Ser., vol. 61, vi + 237 + 5 pp., 4 figs., pls. 47–94.

Rocabert, Lluis. 1934. Contribucio al coneixemant de la fauna ictiologica terciavia catalana. Bul. Inst. Catalana Hist. Nat. 34, nos. 1–5, pp. 78–107, 4 pls.

Roemer, F. 1849A. Texas: Mit besonderer Rücksicht auf deutsche Auswanderung und die physischen Verhältnisse des Landes. Mit einem naturwissenschaftlichen Anhänge und einer topographisch-geognostischen Karte von Texas. Bonn, xv + 464 pp.

———. 1852A. Die Kreidebildungen von Texas und ihre organischen Einschlüsse. Mit einem die Beschreibung von Versteinerungen aus paläozoischen und tertiären Schichten enthaltenden Anhänge und mit von C. Hohe nach der Natur auf Stein gezeichneten Tafeln. Bonn, vii + 100 pp.

Romer, Alfred S. 1945A. Vertebrate paleontology, 2nd ed. Chicago, 687 pp.

———. 1966. Vertebrate paleontology, 3rd ed. Chicago, Univ. Chicago Press, ix + 468 pp., 443 figs., 4 tabs.

Russell, Dale. 1967. Systematics and morphology of American mosasaurs. Yale Peabody Mus. Nat. Hist. Bull. 23, 237 pp.

———. 1970. The vertebrate fauna of the Selma Formation of Alabama. Pt. 7, The Mosasaurs. Fieldiana: Geology Memoirs, vol. 3, no. 7, pp. 365–380.

Savage, T. E. 1902A. Geology of Henry County. Iowa Geol. Surv. 12, pp. 239–302, map, text-figs. 44–55.

Schevill, W. E. 1932. Fossil types of fishes, amphibians, reptiles and birds in the Museum of Comparative Zoology. Mus. Comp. Zool. Bull., vol. 74, no. 4, pp. 59–105.

Schmidt, K. P. 1940. A new turtle of the genus *Podocnemis* from the Cretaceous of Arkansas. Field Mus. Nat. Hist., Geol. Ser. 8, no. 1, pp. 1–12, figs. 1–5.

Serralheiro, A. M. R. 1954. Contribuicao para o conhecimento da fauna ictiologia do Miocenico marinhode Portugal continental. Rev. Lac. Cien Lisbon, ser. 2, vol. 100, pt. 4, pp. 39–119, 2 figs., 4 pls., 5 tabs.

Siler, W. L. 1964. A middle Eocene sirenian in Alabama. Jour. Paleont., vol. 38, no. 6.

Silva Santos, R. da, and Travassos, H. 1960. Contribuicao a paleontologia do estado do Para, piexes fosseis da formacao Probos. Mon. Dept. Nac. Prod. Min., Div. Geol. Min., Brazil, no. 16, ix + 35 pp., 11 figs., 5 pls. (English summary).

Simpson, George G. 1930H. Tertiary land mammals of Florida. Bull. Amer. Mus. Nat. Hist., vol. 59, pp. 149–211, 31 figs.

Slaughter, B. H., and Steiner, M. 1968. Notes on rostral teeth of ganopristine sawfishes, with special reference to Texas material. Jour. Paleont., vol. 42, pp. 223–239, 4 figs.

Slijper, E. J. 1962. Whales. London: Hutchinson of London, 475 pp., 229 figs., Trans. A. G. Pomerans from *Walvissen*, 1958, Amsterdam, D. B. Centen's Uitgeversmoatschappy.

Smith, Burnett. 1909A. Note on the Miocene drum fish—*Pogonias multidentatus* Cope. Am. Jour. Sci., ser. 4, vol. 28, pp. 275–282, 13 text-figs.

Spellman, W. 1863. Catalogue of the Tombigbee Greensand. Report of Geology and Agriculture of the State of Mississippi (for 1860), p. 389.

Springer, Victor G. 1964. A revision of the carcharhinid shark genera *Scoliodon*, *Loxodon* and *Rhizoprionodon*. Proc. U.S. Nat. Mus., vol. 115, pp. 559–632, 14 figs., 17 tabs.

Stewart, A. 1898D. A preliminary description of seven new species of fish from the Cretaceous of Kansas. Kans. Univ. Quart., vii, A, pp. 191–196, pl. 17, text-figs. 1–3.

———. 1899B. *Pachyrhizodus minimus*, a new species of fish from the Cretaceous of Kansas. Kans. Univ. Quart., viii, A, pp. 37–38, 1 text-fig.

———. 1900A. Teleosts of the Upper Cretaceous. Univ. Geol. Surv. Kans., vol. 6, pp. 257–403, 6 tabs., pls. 33–73.

Stirton, R. A. 1932E. A new genus of Artiodactyla from the Clarendon Lower Pliocene of Texas. Univ. Calif. Publ. Bull. Dept. Geol. Sci., vol. 21, pp. 147–168, 3 figs., 5 pls.

Stromer, E. 1903D. Zeuglodon—Reste aus dem oberen Mittdeöcan des Fajum. Zeitr. Pal. Geol. Oesterr.-Ung., vol. 15, pp. 65–100, pls. 8–11, 1 tab.

———. 1905E. Die Fischreste des mittleren und oberen Eocäns von Ägypten. No. 1. Teil. Die Selachii; B., Squaloiden, und 11. Teil. Teleostomi, A. Ganoidei. Beitrage Pal. Geol. Oesterr.-Ung., vol. 18, pp. 163–192, pls. 15, 16.

————. 1921. Untersuchung der Huftbeine und Huftzelenke von Serenia und Archaeocete (mit 6 Textfiguren). Sitzungeber. math.-phys. Klasse der Bayer. Akad. Wiss. zu München, 1921, H.I., pp. 41–60.

Theobold, N. 1934. Contribution à la paléontologie du bassein Oligocène du Haut-Rhin et du territoire de Belfort. Les poissons Oligocènes. Bull. Serv. Carte Geol. Alsace, vol. 2, pp. 117–162, 6 figs., 5 pls.

Thevenin, A. 1896. Mosasauriens de la Craie Grise, de Vaux-Eclusier près de Péronne (Somme). Soc. Geol. France Bull., ser. 3, vol. 24, pp. 900–916.

Thurmond, John T. 1968. A new polycotylid plesiosaur from the Lake Waco Formation (Cenomanian) of Texas. Jour. Paleont., vol. 42, pp. 1289–1296, 4 figs., 3 tabs.

————. 1969. Notes on mosasaurs from Texas. Texas Jour. Sci., vol. 21, pp. 69–80, 2 figs.

————. 1972. Cartilaginous fishes of Trinity Group and related rocks (Lower Cretaceous) of north-central Texas. Southeastern Geology, vol. 13, pp. 207–227, 14 text-figs.

————. 1974. Lower vertebrate faunas of the Trinity Division in north-central Texas. Geoscience and Man, vol. 8, pp. 103–129, 1 pl., 15 text-figs.

True, F. W. 1908. The Fossil Cetacean, *Dorudon serratus* Gibbes. Bull. Mus. Comp. Zool. Harvard, vol. 52, no. 4, pp. 65–78.

Tuomey, Michael. 1847A. Discovery of the cranium of the zeuglodon. Am. Jour. Sci., ser. 2, vol. 4, pp. 283–285, 1 text-fig.

————. 1847B. Notice of the discovery of a cranium of the zeuglodon. Proc. Acad. Nat. Sci. Philadelphia, vol. 3, pp. 151–153, 2 woodcuts.

————. 1847C. Notice of the discovery of a cranium of the zeuglodon *(Basilosaurus)*. Jour. Acad. Nat. Sci. Philadelphia, ser. 2, vol. 1, pp. 16–17.

————. 1848A. Report on the geology of South Carolina. Columbia, S.C., vi + 293 + lvi pp., geol. map, 47 text-figs.

————. 1850A. Remarks on a fossil reptile from Alabama, belonging to the genus *Leiodon*. Followed by remarks by Prof. L. Agassiz. Proc. Am. Assoc. Adv. Sci., 3rd meeting, Charleston, S.C., 1850, p. 74.

————. 1850B. First biennial report of the geology of Alabama. Tuskaloosa, xxxii + 176 pp.

————. 1858A. Descriptions and figures of *Ctenacanthus elegans, Cladodus newmani, C. magnificus*. Second Biennial Rept. Geol. Ala., xix + 292 pp.

Uyeno, T. 1961. Late Cenozoic cyprinid fishes from Idaho with notes on other fossil minnows of North America. Papers Mich. Acad. Sci., vol. 46 (1960), pp. 329–344, 3 figs.

————, and Miller, R. H. 1963. Summary of Late Cenozoic freshwater-fish records for North America. Occ. Paper Mus. Zool. Univ. Mich., no. 631, pp. 1–34.

VanderGeyn, W. A. G. 1937B. Das Tertiär der Niederlande mit besonderer Berücksichtigung der Selachierfauna: Leidische geologische mededeelingen, vol. 9, pp. 117–361.

Vickers, E. D. 1962. Notes on Miocene mammal remains in the Georgia-Florida district. Ga. Min. Newsletter, vol. 15, pp. 28–29, 1 fig.

Wallace, W. D. 1963. Alabama boneyard sheds new light on whales. Sci. Digest, vol. 53, no. 4, pp. 77–80, 4 figs.

Warren, J. C. 1855. The *Mastodon giganteus* of North America. 2nd ed., Boston, 260 pp., 4 figs., 31 pls., p. 162, pl. 28, fig. A.

Weeland, G. R. 1900B. Some observations of certain well-marked stages in the evolution of the testudinate humerus. Am. Jour. Sci., ser. 4, vol. 9, pp. 413–424, 23 text-figs.

Welles, S. P. 1943. Elasmosaurid plesiosaurs with description of new material from California and Colorado. Mem. Univ. Calif. 13, pp. 125–254, front., 37 figs., 18 pls.

———. 1952. A review of the North American Cretaceous elasmosaurs. Univ. Calif. Publ. Geol. Sci., vol. 29, pp. 47–144, 25 figs.

———. 1962. A new species of elasmosaur from the Aptian of Columbia and a review of the Cretaceous plesiosaurs. Univ. Calif. Pub. Geol. Sci., vol. 44, 89 pp., 23 figs., 4 pls.

Wesler, W. 1935D. Die Fischreste aus dim Budaer (Ofner) Mergel des Gellerthegy (Blocksburges) bei Budapest. Am. Hist.-Nat. Mus. Nation. Hugarici, 29 (pas. mun. geol. pal.), pp. 29–34, 6 figs.

Wetmore, Alexander. 1962. Notes on fossil and subfossil birds. Smiths. Coll., vol. 145, no. 2, pp. 1–17, 2 figs.

White, E. I. 1926A. Eocene fishes from Nigeria. Geol. Surv. Nigeria, Bull. 10, pp. 1–82, pls. 1–18, figs. 1–6.

———. 1931. The vertebrate faunas of the English Eocene. I: From the Thanet Sands to the Basement Bed of the London Clay. Brit. Mus. (Nat. Hist.), pp. 1–121, 1 pl.

———. 1956. Eocene fishes of Alabama. Bull. Am. Paleont., vol. 36, no. 156, pp. 122–152, 1 pl., 97 text-figs.

White, T. E. 1940. Holotype of *Plesiosaurus longirostris* Blake and classification of the plesiosaurs. Jour. Paleont., vol. 14, pp. 451–467.

Williston, S. W. 1898. Mosasaurs. Univ. Geol. Surv. Kansas, vol. 4, pp. 83–221.

———. 1900A. Some fish teeth from the Kansas Cretaceous. Kansas Univ. Quart., vol. 9, pp. 27–42, pls. 6–14.

———. 1903A. North American plesiosaurs. Pt. I, Field Columbian Mus. Geol., vol. 2, pp. 1–73, pls. 1–29.

———. 1906B. North American plesiosaurs: *Elasmosaurus*, *Cimoliasaurus*, and *Polycotylus*. Am. Jour. Sci., ser. 4, vol. 9, pp. 221–236, pls. 1–6, text-figs. 1–5.

———. 1907A. The skull of *Brachauchenius*, with observations on the relationships of the plesiosaurs. Proc. U.S. Nat. Mus. 32, pp. 477–489, pls. 34–37.

———. 1908C. North American plesiosaurs: *Trinacromerum*. Jour. Geol., vol. 16, pp. 715–736, figs. 1–15.

———. 1908E. The evolution and distribution of the plesiosaurs. Science, n.s., vol. 27, pp. 726–727.

Wiman, C. 1920. Some reptiles from the Niobrara Group in Kansas. Geol. Inst. Uppsala Bull., vol. 18, pp. 9–18.

Winge, H., and Miller, G. S. 1921A. A review of the interrelationships of the Cetacea. Smiths. Misc. Coll., vol. 72, no. 8, pp. 1–97.

Winkler, T. C. 1874B. Mémoire sur des dents de poissons du terrain bruxellian. Arch. Mus. Teyler, vol. 3, pp. 295–304, pl. 7.

———. 1876A. Mémoire sur quelques restes de poissons du système heersien. Arch. Mus. Teyler, vol. 4, pp. 1–15, pls. 1–2.

———. 1876B. Deuxième mémoire sur des dents de poissons fossiles du terrain bruxellian. Arch. Mus. Teyler, vol. 4, pp. 16–48, figs. A–F, pl. 2.

Wixson, B. E. 1963. Fossil shark teeth in the study of Cretaceous environments near Terlingua, Texas. Texas Jour. Sci., vol. 15, pp. 415–416 (abs.).

Woodward, A. S. 1889D. Catalogue of the fossil fishes in the British Museum. Pt. 1. Containing the *Elasmobranchii*. London, 1889, pt. 1, xlvii + 474 pp., pls. 1–17, 13 woodcuts.

———. 1891A. Catalogue of fossil fishes in the British Museum. Pt. 2. Containing the *Elasmobranchii (Acanthodii), Holocephali, Ichthyodorulites, Ostracodermi, Dipnoi,* and *Teleostomi (Crossopterygii),* and chondrostean *Actinopterygii,* xliv + 567 pp., 16 pls., 57 text -figs.

Wyman, J. 1845. Communication on the skeleton of *Hydrarchos sillimani.* Proc. Boston Soc. Nat. Hist., vol. 2, pp. 65–68.

———. 1850. Notice of remains of vertebrated animals found at Richmond, Virginia. Am. Jour. Sci., ser. 2, vol. 10, pp. 228–235, woodcut.

Zangerl, Rainer. 1948A. The vertebrate fauna of the Selma Formation of Alabama. Pt. I, Introduction. Fieldiana: Geology Memoirs 3 (1 and 2), pp. 1–56, 16 figs., 4 pls.

———. 1948B. The vertebrate fauna of the Selma Formation of Alabama. Pt. II, The pleurodiran turtles. Fieldiana: Geology Memoirs 3 (1 and 2), pp. 1–56, 16 figs., 4 pls.

———. 1953A. The vertebrate fauna of the Selma Formation of Alabama. Pt. II, The turtles of the family Protostegidae. Fieldiana: Geology Memoirs 3 (3 and 4), pp. 55–277.

———. 1953B. The vertebrate fauna of the Selma Formation of Alabama. Pt. IV, The turtles of the family Toxochelyidae. Fieldiana: Geology Memoirs 3 (3 and 4), pp. 55–277.

———. 1960. The vertebrate fauna of the Selma Formation of Alabama. Pt. V, An advanced cheloniid sea turtle. Fieldiana: Geology Memoirs 3, pp. 279–312, figs. 125–145, pls. 30–33.

Index